SpringerBriefs in Food, Health, and Nutrition

Springer Briefs in Food, Health, and Nutrition present concise summaries of cutting edge research and practical applications across a wide range of topics related to the field of food science.

For further volumes:
http://www.springer.com/series/10203

William Aspray • George Royer
Melissa G. Ocepek

Formal and Informal Approaches to Food Policy

Springer

William Aspray
School of Information
The University of Texas at Austin
Austin, TX, USA

George Royer
Austin, TX, USA

Melissa G. Ocepek
The University of Texas at Austin
Austin, TX, USA

ISSN 2197-571X ISSN 2197-5728 (electronic)
ISBN 978-3-319-04965-6 ISBN 978-3-319-04966-3 (eBook)
DOI 10.1007/978-3-319-04966-3
Springer Cham Heidelberg New York Dordrecht London

Library of Congress Control Number: 2014932542

Printed on acid-free paper

Springer is part of Springer Science+Business Media (www.springer.com)

Acknowledgements

This book is a product of the Informed Stomach Research Group in the School of Information at the University of Texas at Austin. We thank the School of Information for its infrastructural support in carrying out this and other research on food and information in the United States. William Aspray was supported in this research in part by research funds associated with the Bill and Lewis Suit Professorship in Information Technologies and the STARS Plus Fund provided by the University of Texas at Austin. Over the past 5 years, the Informed Stomach Research Group has received intellectual guidance from Harrison Archer, Zak Archer, Lecia Barker, Andrew Dillon, Lorrie Dong, Philip Doty, Elizabeth Englehardt, Nathan Ensmenger, Kenneth Fleischmann, Patricia Galloway, Barbara Hayes, Steve Mannheimer, Susan Smulyan, and Yan Zhang. Research assistance in archives, literature review and retrieval, and copyediting have been provided during this same period by Lynn Eaton, Kip Keller, and Elliot Williams. We also appreciate the extensive comments provided by seven anonymous reviewers who read one of Chaps. 3, 4, or 5. Thanks to all.

Contents

Chapter 1
Introduction

This brief book presents an overview, together with four case studies, of formal and informal approaches to food policy in the United States. Chapter 2 describes the many different kinds of actors and techniques used in formal and informal approaches to food policy. Formal approaches are, by definition, those taken by all branches (executive, legislative, and judicial) of federal, state, and local government. In contrast, informal approaches are those taken by companies and industries, scientific and professional organizations, public interest groups, protest groups, celebrities, and individuals.

Chapters 3 and 4 are concerned with the issue of childhood obesity, which by the year 2000 had reached epidemic proportions in the United States. Chapter 3 presents a history of formal and informal means to regulate the advertisement to children of nonnutritious foods on television and the Internet. Chapter 4 offers a history of formal and informal means to improve healthy eating in the public schools, both through the supply of more nutritious government-subsidized breakfasts and lunches in the schools and the regulation of less healthy competitive foods available in vending machines and school stores.

Chapters 5 and 6 present two cases studies of a particular kind of informal approach to food policy, the use of the bully pulpit of the White House by First Ladies Eleanor Roosevelt and Michelle Obama to improve healthy eating and physical fitness of all Americans. The informal approaches of these two First Ladies are told in the context of the formal approaches to policy by their husbands' administrations in the areas of agriculture, physical fitness, and nutrition policy.

While there are connections between these chapters, each chapter can be read on its own. Chapters 3 through 6 illustrate the general comments in Chap. 2. Chapters 3 and 4 are connected to Chaps. 5 and 6 in several ways. One section of Chap. 4 provides another example of the use of the bully pulpit—by former President Clinton rather than by a First Lady—to improve self-regulation by the beverage industry of access to sweetened drinks in schools. Much of Chap. 6 is focused on efforts by Michelle Obama to fight childhood obesity (improved access to nutritious

W. Aspray et al., *Formal and Informal Approaches to Food Policy*, SpringerBriefs in Food, Health, and Nutrition, DOI 10.1007/978-3-319-04966-3_1,

foods, improved food education, and increased physical fitness for children)—but somewhat in different ways from what is covered in Chaps. 3 and 4.

The final two sections of this Introduction offer background information for the remainder of the book. The next section provides an overview of childhood obesity in support of the policy histories given in Chaps. 3 and 4. The final section of this Introduction supports Chaps. 5 and 6 by providing background information on the history of the bully pulpit as an informal means of persuasion, especially as it has been employed by First Ladies throughout American history.

1.1 Childhood Obesity in America

Chapters 3 and 4 discuss two important factors in the rise of childhood obesity, advertising of unhealthy foods on television and the Internet, and unhealthy food available in public schools, respectively. In this section, we provide some information on childhood obesity in America that will serve as background for the discussions in Chaps. 3 and 4.

The problem of childhood obesity had been growing in the 1980s and 1990s, but it only received national attention in the new century. In 2001 David Satcher, Surgeon General of the United States, announced that obesity had reached epidemic proportions among America's children, adolescents, and adults (Satcher 2001). He noted that since 1980 the percentage of children and adolescents who are overweight had tripled. Prevalence of obesity among all American children had risen from 5.1 % in 1971–1974, to 10.0 % in 1988–1994, to 15.4 % in 2001–2002, according to the National Health and Nutrition Exam Survey. The percentage of children who are obese has flattened out for children overall but it has continued to increase for boys aged 12–19. The numbers are much higher for minorities, e.g., 24.8 % for non-Hispanic blacks in 2009–2010 (Fryar et al. 2012; Ogden et al. 2010). Adults are classed as being overweight if their body mass index (BMI) is greater than 25 and obese if their BMI is greater than 30. There are slight variations in the definitions used by different organizations to measure childhood obesity but those differences make little difference in policy discussions (Harvard School of Public Health 2013; also see Mayo Clinic Staff 2012).

There are numerous health impacts of childhood obesity—both immediate and long term. Immediate impacts include cardiovascular disease including high cholesterol and high blood pressure, prediabetes (a high risk of developing diabetes), bone and joint problems, sleep apnea, and social and psychological problems including stigmatization and low self-esteem. Long-term risks include increased likelihood of the health problems associated with adult obesity, heart disease, Type 2 diabetes, stroke, osteoarthritis, and various types of cancer (breast, colon, endometrium, esophagus, kidney, pancreas, gall bladder, thyroid, ovary, cervix, prostate, multiple myeloma, and Hodgkin's lymphoma). (Centers for Disease Control 2013).

The likelihood of obesity in adulthood for children who are obese is high. In a literature review that examined 17 published reports published between 1970 and

1992 (based on 15 study populations), 26–41 % of obese preschool children were obese as adults and 42–63 % of obese school-age children were obese as adults (Serdula et al. 1993). In another longitudinal study of 277 white males and 278 white females, being overweight at age 35 could be predicted with high levels of certainty for participants who were overweight at age 18, with good levels of certainty for participants who were overweight at age 13, and with moderate certainty for participants under 13 years of age (Guo and Chumlea 1999; also see Whitaker et al. 1997).

Despite the Surgeon General's call for action, the problem persisted through the first decade of the new century, although increasing at a slower pace than in the 1980s and 1990s. As of 2012, 17 % of American children and adolescents were obese, and another 22 % were overweight.

Obesity has a significant economic and labor impact. On average, obese employees miss more days from work due to short-term absences, long-term disability, and premature death than nonobese employees. They may also work at less than full capacity when at work. Data from the Medical Expenditure Panel Survey for 2000–2005 show that per capita spending on annual medical costs is $2,741 (in 2005 dollars) more for obese individuals than for nonobese individuals (Cawley and Meyerhoefer 2012). This study also showed that, in 2005, 21 % of medical spending ($190 billion) could be attributed to obesity. Another study, using data from the National Health Expenditure Accounts Dataset, calculated that obesity costs $147 billion in health care spending in the United States in 2006 (Finkelstein et al. 2009).

Widespread obesity has many negative economic consequences for the nation. It drives up the health and life insurance rates for people who are not overweight as well as for those who are, and it raises serious issues about national productivity and the labor force. Impact on the labor force includes the military labor force. Early in the Obama presidency, past military leaders pointed to obesity as a national security issue (Mission: Readiness 2010). This heightened the resolve of both Michelle and Barack Obama to address obesity as a national problem.

Although the reason for weight gain is simple—more calories consumed in food than burned in activity—the causes are complicated and environmental. Part of the problem has to do with the amount of physical activity that Americans get. But the larger part of the problem has to do with what Americans eat and in what quantities. This is a complex issue that involves many dimensions such as educational levels of different demographic groups, the socioeconomics of housing and access to food, cultural differences in diets among different ethnic populations in the United States, and many more issues. We return to these issues in Chaps. 3, 4, and 6.

1.2 The Historical Use of the Bully Pulpit by Presidents and First Ladies

In addition to the formal means of the federal government, the Office of the President has an important informal way to shape the behavior of individuals and the businesses that serve them. This is the bully pulpit, the use of the high profile and nationally respected office of the presidency to make speeches and other symbolic acts to persuade people and organizations to behave in certain ways for the good of the country. This section provides information about the bully pulpit, especially as it has been employed by First Ladies, as background for the discussions of its use in the food realm by Eleanor Roosevelt and Michelle Obama.

The president is empowered to speak in a way that no other individual is. Unlike members of Congress or the Supreme Court, the president is elected by the entire voting population. The content of the president's speeches does not require authorization by Congress or approval by the courts. The president's ability to take advantage of the bully pulpit is limited only by his (only male up until now) speech-writing and speech-giving ability. Does the president have the ability to find good metaphors to persuade the populace to think about issues in a certain way and to act in accordance with these beliefs? Sometimes, the bully pulpit is intended to get the citizens to act and think in certain general ways; at other times it is intended to get the public to support a particular legislative agenda. The bully pulpit can be a powerful tool for the president against Congress, the courts, bureaucracy, political parties, and special interest groups.

As a lead speechwriter for President Ronald Reagan, Peggy Noonan was aware of the importance of the bully pulpit and became curious about its history. As a result of her investigations, she wrote: "You know the derivation of the phrase 'bully pulpit'? Teddy Roosevelt invented it. It's one of his formulations. 'Bully' was one of his favorite adjectives—it was his favorite favorable adjective. It meant terrific. 'Bully pulpit' meant a place from which one could influence more minds than from any other lectern" (as quoted in Muir 1992).

The bully pulpit and the role of the president as public persuader only came into its own in the twentieth century. In the nineteenth century, presidents other than Abraham Lincoln made few public speeches, and those they did make were largely ceremonial. Teddy Roosevelt argued that there had been two types of president: those such as William Taft and James Buchanan who only acted when the Constitution specifically allows action; and those such as Andrew Jackson and Abraham Lincoln who felt free to act any time the Constitution did not specifically disallow action. Roosevelt saw himself in the latter group (Dorsey 1995; Gelderman 1997; Hagedorn 1926; Muir 1992, 1995).

The change toward an activist presidency began with Theodore Roosevelt's predecessor, William McKinley, who courted public opinion both in his quest for the office and after his arrival in the White House. McKinley made the first campaign film, using Thomas Edison's new film technology just a month after the technology was installed in theaters. In his early days in the White House, McKinley hosted a reception for journalists, which won their favor. He also hired a staff member, George Cortelyou, who served as the first official White House press secretary and who put into place a formal system for handling requests from correspondents. This proved especially useful to the president in 1898, during the Spanish-American War.

Teddy Roosevelt was only 42 years old when he became president in 1901, following McKinley's assassination. He faced a nation in tumult. There was unbridled power of industrial capitalism and mounting corporate greed, strikes by coal miners, growing poverty across America, and international military unrest. Roosevelt believed the presidency should be the center of both policymaking and leadership for the country, not merely the office that administers the programs of Congress, as had been commonplace during most of the nineteenth century. In order to attain this goal, he took steps to shape public opinion, not simply to represent the opinions of the public or of his party. He used the force of his considerable personality to promulgate moral reform in the nation.

Although he was young when he moved to the White House, Roosevelt was already experienced at shaping public image. He had cultivated journalists from the age of 23, when he entered politics as a New York state representative. Later, as the New York City police commissioner, he worked closely with the famous reform-minded reporters Jacob Riis and Lincoln Steffens to his and their mutual benefit. During the Spanish-American War, he convinced *Scribner's* magazine to let him pen a series of articles about his exploits in fighting the Spanish in Cuba. Upon being elected governor of New York in 1898 as a war hero, Roosevelt met twice a day with the press. Once he came to the White House, he managed his public relations by introducing the practices of releasing bad news on Friday afternoons so that it appeared in the little-read Saturday papers, leaking information to reporters as trial balloons on policy, and orchestrating events so that there was ample photographic coverage to assure front-page articles. He worked hard to shape public opinion. For example, through 18 months of use of the bully pulpit, finally in 1906 he persuaded Congress to pass the Hepburn Act. This law, which extended federal regulatory powers over the railroads, was opposed by his own political party. He also used the bully pulpit to educate leaders of large business trusts about their moral responsibilities and leadership role in society, and the need for enlightened self-regulation of their companies. His bully pulpit also served to dampen the impact of muckraking journalists on public opinion and limit antibusiness sentiment among the general public—in order to reduce the tendency toward socialist solutions by labor unions, most notably in the meatpacking industry.

Although the main interest here is in Teddy Roosevelt's role in creating the bully pulpit, as an aside we note that he did use it to promote good health. He was an advocate and icon of vigorous health. He practiced and endorsed Fletcherism, Horace Fletcher's popular dietary regime based on the practice of good health through eating less and chewing more. He registered as a patient at John Harvey Kellogg's Battle Creek Sanitarium in Michigan, run by the Seventh Day Adventist church, which advocated vegetarianism, abstinence from alcohol, and exercise. The sanitarium attracted a wide celebrity clientele, including President William Taft, Tarzan actor Johnny Weissmuller, Henry Ford, and Thomas Edison. After his presidency, in 1913, Teddy gave speeches in favor of a public school lunch program as a way to fight childhood malnutrition. He promoted a national program to revive the buffalo herds of North America and promoted the healthful virtues of buffalo meat. It was, however, too often a case of "do as I say, not as I do" with Teddy. He did not always choose to eat in the most healthy way—a typical breakfast on his ranch was a cup of coffee, a few mouthfuls of bread, and some elk jerky.

President Woodrow Wilson and his successors continued the movement toward a persuasive presidency. Wilson was the first president in a 100 years to speak directly to both houses of Congress. He kept his 1913 speech to the joint houses of Congress on tariff reform short to ensure that it would be printed in its entirety in newspapers across the country. Presidents Warren Harding and Calvin Coolidge were the first to employ ghostwriters, which enabled an increase in the number of presidential public addresses given. The fact that ghostwriters were used was kept secret during their presidencies, but it became acceptable to make this fact known to the public by the time of Franklin Delano Roosevelt's presidency. In preparing for war, FDR effectively used his fireside chats on the radio in the years 1937–1940 to move the public from a position of isolationism to one of internationalism. His style differed from that of his distant cousin Teddy: more one of calmly explaining problems and how his administration was going to handle them rather than exhorting the public. The fireside chats were notable because of Roosevelt's soft delivery, which was in stark contrast to his booming campaign speeches. Several listeners would write to Roosevelt after a fireside chat and remark on the intimacy with which he spoke—remarking that it was like a friend had stopped by to chat and that his words gave them comfort during the hard times following the depression and leading up to war (Lenthall 2007; Ryan 1995).

John Kennedy and Ronald Reagan effectively used the bully pulpit. Some other presidents were less effective, notably Dwight Eisenhower, Jimmy Carter, and George H.W. Bush. President Eisenhower, in particular, was not interested in the persuasive values of the bully pulpit. When told that one of his speeches was not eloquent, he responded, "If good writing was necessary for good leading, the country ought to turn to Hemingway" (Quoted in Gelderman 1997). With the rise of cable television, there was a downward spiral of television coverage of presidential speeches. Fewer people watched major presidential addresses since they had more programming choices, reducing the impact of the speeches. The lower television ratings of presidential addresses in turn caused the networks to cover even fewer presidential speeches, which in turn caused presidents to decrease the number of major public addresses they made. (On the use of the bully pulpit by Theodore Roosevelt's successors, see Fournier 2011; Gelderman 1997; Greenberg 2011; Lynn 2009; Muir 1995; Young and Perkins 2005.)

The high profile of the presidency also extends a bully pulpit to former presidents, vice presidents, and First Ladies—although always in an attenuated way compared to the opportunities offered to the sitting president. The focus in Chaps. 5 and 6 is on the use of the bully pulpit by First Ladies Eleanor Roosevelt and Michelle Obama to speak out on food and nutrition. Sometimes their work was largely independent of the formal policy work of their husbands, as in the demonstration garden in the Obama White House. Other times the work was carried through in close collaboration with the president, as in the physical fitness and school lunch programs in the Obama administration. Sometimes the efforts of the First Lady went beyond what the president was politically comfortable in endorsing, as in the case of Eleanor Roosevelt's support for planned homestead communities such as the one in Arthurdale, West Virginia.

The First Lady, i.e., the wife of the president or the person who serves as the White House hostess when the president is unmarried or his wife is unable to carry out these duties, holds a position of high profile and esteem. However, it is also a post that carries many restrictions and sensitivities. The First Lady is expected to conform to society's existing gender roles. She must serve as an unpaid hostess to visiting dignitaries in order to advance the president's and the nation's goals. Great tact is needed to avoid creating enemies for the president. Great discretion must be used to avoid revealing state secrets. The First Lady must not overshadow her husband. Historically, she could not be seen by the public as overstepping her role as a helpmeet and moral sounding board, and in particular not be seen as interfering in the governing of the nation (as Edith Wilson was regarded as doing when Woodrow Wilson was ill near the end of his presidency).

It takes great effort and dexterity, as well as luck, for a First Lady to find her own voice and be effective in carrying out a mission while in the White House. For example, Lou Henry Hoover was an extremely able individual. She had a Stanford University geology degree, translated a book on metals from Latin, bicycled through China during the Boxer Rebellion, and organized a committee to repatriate 10,000 Americans living in Europe during the First World War. But she was never able to break free of the conventional chains and become an effective First Lady (Beasley 2010). The model of the successfully activist First Lady entering the twentieth century was Dolley Madison, who not only had social skills and political savvy to help her husband, but took a "woman's" interest in the treatment of orphans (National First Ladies' Library n.d.; Allgor 2012). Eleanor Roosevelt, who held the post longer than anyone else, replaced Dolley Madison as the iconic First Lady.

Even after Roosevelt's path-breaking efforts, activism by First Ladies continued to conform with society's notions of proper gender roles and women's interests. This is seen in the principal missions adopted by the First Ladies: Jackie Kennedy (historic preservation of the White House), Lady Bird Johnson (environmental protection and highway beautification), Pat Nixon (volunteerism), Betty Ford (women's rights), Rosalyn Carter (mental illness), Nancy Reagan (drug awareness), Barbara Bush (literacy), Hillary Clinton (healthcare), and Laura Bush (childhood literacy). Food and children's health fell squarely within this feminine domestic sphere in American public opinion.

References

Allgor C (2012) Dolley Madison: The problem of national unity. Westview Press, Boulder, CO

Beasley MH (2010) Eleanor Roosevelt: Transformative first lady. University of Kansas Press, Lawrence

Cawley J, Meyerhoefer C (2012) The medical care costs of obesity: An instrumental variables approach. J of Health Econ 31:219–230

Centers for Disease Control and Prevention (2013) Childhood obesity facts. http://www.cdc.gov/healthyyouth/obesity/facts.htm. Accessed 27 Aug 2013

Dorsey LG (1995) The Frontier Myth in Presidential rhetoric. Western J. Comm. 59:1–19

Finkelstein EA, Trogdon JG, Cohen JW, Dietz W (2009) Annual medical spending attributable to obesity: payer- and service-specific estimates. Health Aff (Millwood) 28:w822–w831
Fournier R (2011) No TR: The limits of Obama's bully pulpit. Decoded (blog), National Journal. http://decoded.nationaljournal.com/2011/12/obama.php. 6 Dec 2011, Accessed 16 Dec 2013
Fryar CD, Carroll MD, Ogden, CL (2012) Prevalence of obesity among children and adolescents: United States, trends 1963–1965 through 2009–2010. Centers for Centers for Disease Control and Prevention. http://www.cdc.gov/nchs/data/hestat/obesity_child_09_10/obesity_child_09_10.htm. Accessed 27 Aug 2013
Gelderman C (1997) All the Presidents' words: The bully pulpit and the creation of the virtual presidency. Walker, New York
Greenberg D (2011) Beyond the bully pulpit. Wilson Q 35(3):22–29
Guo SS, Chumlea WC (1999) Tracking of body mass index in children in relation to overweight in adulthood. Am J Clin Nutr 70:145s–148s
Hagedorn H (ed) (1926) Works of Theodore Roosevelt. Charles Scribner's Sons, New York
Harvard School of Public Health (2013) Defining childhood obesity. The Obesity Prevention Source. http://www.hsph.harvard.edu/obesity-prevention-source/obesity-definition/defining-childhood-obesity/#References. Accessed 4 Sept 2013
Lenthall B (2007) Radio's America: The Great Depression and the rise of modern mass culture. University of Chicago Press, Chicago
Lynn LE Jr (2009) Theodore Roosevelt redux: Barack Obama confronts American bureaucracy. Int J Public Adm 32(9):773–780
Mayo Clinic Staff (2012) Childhood obesity. MayoClinic.com. http://www.mayoclinic.com/health/childhood-obesity/DS00698. 4 May 2012, Accessed 27 Aug 2013
Mission: Readiness (2010) Too fat to fight: Retired military leaders want junk food out of America's schools. Mission: Readiness, Washington, DC
Muir WK Jr (1992) The bully pulpit: The Presidential leadership of Ronald Reagan. Institute for Contemporary Studies, ICS Press, San Francisco
Muir WK Jr (1995) The bully pulpit. Pres Stud Q 25(1):13–17
Ogden CL, Carroll MD, Curtin LR, Lamb MM, Flegal KM (2010) Prevalence of high body mass index in US children and adolescents, 2007–2008. J Am Med Assoc 303:242–249
Ryan H (1995) Franklin Delano Roosevelt. In: Ryan H (ed) U.S. Presidents as orators: A bio-critical sourcebook. Greenwood Press, Westport, p 146–167
Satcher D (2001) Surgeon General's call to action to prevent and decrease overweight and obesity. US Department of Health and Human Services. http://www.surgeongeneral.gov/topics/obesity/. Accessed 16 Dec 2013
Serdula MK, Ivery D, Coates RJ, Freedman DS, Williamson DF, Byers T (1993) Do obese children become obese adults? A review of the literature. Prev Med 22(2):167–177
The National First Ladies' Library (nd) First lady biography: Dolley Madison. http://www.firstladies.org/biographies/firstladies.aspx?biography=4. Accessed 22 Aug 2013
Whitaker RC, Wright JA, Pepe MS, Seidel KD, Dietz WH (1997) Predicting obesity in young adulthood from childhood and parental obesity. N Engl J Med 37:869–873
Young G, Perkins WB (2005) Presidential rhetoric, the public agenda, and the end of Presidential television's 'golden age.' J Pol 67(4):1190–1205

Chapter 2
Formal and Informal Approaches to Food Policy

Some in industry may criticize us for using our bully pulpit to encourage companies to do a better job of marketing healthier products to youth. Such criticism would be misguided...A little government involvement—combined with a lot of private sector commitment—can go a long way toward the healthier future for our children that all of us want to see.

(FTC Commissioner Jon Leibowitz, Leibowitz 2008 as quoted in Mello 2010)

2.1 Overview

In the preamble to the U.S. Constitution, the federal government is charged with taking care of the general welfare of the population. One of the population's most basic needs is adequate nutrition. Thus, the federal government has long had a mandate to act on food policy. The first U.S. president, George Washington, who was a large landowner and agricultural experimenter (unsuccessfully) proposed a National Board of Agriculture to Congress. The third president, Thomas Jefferson, who owned a large plantation in Virginia and was a leading agricultural scientist, inventor, and breeder, supported agrarian self-sufficiency as the bedrock of the nation's economy and thus supported low tariffs so that farmers would not pay too much for the supplies and tools they needed. The federal government's first active intervention in agriculture began in the 1830s, through the efforts of Henry Leavitt Ellsworth, a lawyer and farmer who also served as the president of Aetna Insurance Company. Upon being appointed Commissioner of Patents by President Andrew Jackson, Ellsworth began to collect new varieties of seeds and plants from across the nation and agitated Congress to provide funds to support agricultural interests. In 1839 an Agricultural Division was created within the Patent Office and charged with collecting agricultural statistics. The U.S. Department of Agriculture (USDA) was founded in 1862 as a direct descendent of this Patent Office activity, and it has continued to

W. Aspray et al., *Formal and Informal Approaches to Food Policy*, SpringerBriefs in Food, Health, and Nutrition, DOI 10.1007/978-3-319-04966-3_2,

carry out various research, educational, and regulatory activities up until the present time (Waggoner 1976; National Agricultural Hall of Fame n.d.).

Food policy is complex. It involves many different issues and many different interest groups. Consider, for example, the case of childhood obesity. There are many different players. There is the health care community, including individual doctors, health care provider organizations, insurance companies, public health officials, and public health advocates. While all of these players are interested in healthier children, there are disagreements within this community over costs and responsibilities; for example, individual doctors and public health officials might desire certain types of actions that provider organizations or insurance companies might balk at providing for financial reasons. Another player is the food industry, ranging from farmers, to grocers, food manufacturers, and restaurants—together with their various industry and trade organizations. The food industry is not monolithic in its policy attitudes and actions toward childhood obesity; for example, fast-food restaurants may take positions oppositional to organic farmers. Local, state, and federal governments are torn by conflicts between their interest in public health and their support of various industries, not only including the food industry but also the advertising, media, construction, and transportation industries on such questions as taxation of unhealthy products, First Amendment rights to free commercial speech, and the regulation of the built environment to make cities walking- and bike-friendly. Children spend many hours in school, and principals and school boards have to live with the tension between access to nonhealthy foods in the schools and the loss of revenue from programs sponsored by the food and beverage industries. In order to provide solutions to some of the issues relating to childhood obesity, policymakers must address some of America's thorniest socioeconomic policy issues, such as the lack of access to fresh fruits and vegetables or the underfunded schools in many low-income neighborhoods. Thus, one can see how complex these food policy issues can be.

While we are only able in this brief book to discuss in detail two aspects of the policy fight against childhood obesity—advertising of unhealthy foods to children (on the television and Internet) and unhealthy food in schools—we summarize below the overall policy approach. Before considering policy initiatives, one might ask why the marketplace cannot solve this problem without the intervention of government. Seiders and Petty (2004) identify four market failures related to food choice. They are (using their exact language): lack of disseminated information on the causes and consequences of obesity, the probabilistic and deferred nature of obesity-related harms, lack of accessible and usable nutritional information related to obesity, and the lack of alternative food choices for some consumers.

Three high-level officials at the Centers for Disease Control (Frieden et al. 2010) have provided an excellent overview of formal policy approaches to childhood obesity. Here we draw heavily not only from their account but also from those of several other scholars (Alderman et al. 2007; Seiders and Petty 2004; Anomaly 2012; Schwartz and Brownell 2007; Mello et al. 2006; Sugarman and Sandman 2007; Thaler and Sunstein 2008; Moorman and Price 1989; McGinnis et al. 1999).

Frieden and his colleagues identify three aspects of food policy that can have an impact on childhood obesity. They concern changing the pricing of foods, altering the public's exposure to different kinds of foods, and changing the image of healthy and unhealthy foods. Pricing policy might include, for example, taxing unhealthy foods to make them more expensive than healthy foods or using agriculture and school meal subsidies to encourage the increase in growth and consumption as well as reduction in cost of healthy food products. Exposure policy might involve, for example, policies that encourage an increase in the number of farmer's markets or supermarkets with fresh fruits and vegetables located in the so-called food deserts that occur commonly in low-income, inner city neighborhoods (and in other places as well) or zoning regulations that limit the density or proximity to schools of fast-food restaurants and convenience stores. Image policy is intended to make healthy foods look more attractive and unhealthy foods less so, such as restrictions on food advertisements to children, providing access to nutritional information on menus in restaurants, and counteradvertising that shows the long-term negative health impacts of regular consumption of highly sweetened beverages.

Policies addressed at increasing children's physical activity also have an impact on childhood obesity. Such policies include changing built environment design (creation of parks, wide sidewalks, and bike lanes) to make walking and biking easier, safer, and more attractive; encouraging children to replace sedentary activities such as television watching and video game playing with more active pursuits; and improving physical education in schools both inside and outside of formal school hours (Khan 2011; Perdue et al. 2003; but also see the other sources listed two paragraphs earlier).

The government also carries out this mission of fighting childhood obesity in other formal ways. These include providing funds for the rigorous assessment of community-level interventions intended to address relevant issues, paying for research on the relationship between diet and health, providing expert assessments of the body of scholarship studying the relationships between diet and health, producing data on consumer behavior as it relates to food and nutrition, offering dietary advice to consumers through tools such as food pyramids and educational campaigns, mandating warning labels or nutritional information on food product packaging, and regulating nutrition claims of food producers (Ippolito 1999, but also see the other sources listed three paragraphs earlier).

2.2 Formal Policy

All three branches of the federal government participate in food policy. In the Executive Branch the most important player is the USDA, which, for example, regulates agriculture, conducts research on food and diet, and provides educational tools such as the food pyramid. However, we will show in Chap. 3 how the Federal Trade Commission, the Federal Communication Commission, and Health and Human Services have each played a role in food advertising regulation. Chapter 4 discusses how the Government Accountability Office (formerly the General Accounting Office) has tracked commercialism, including food commercialism, in the public school system.

Pure Food and Drug Act	1906
Meat Inspection Act	1906
Agricultural Adjustment Act	1933
Commodity Credit Corporation Charter Act	1933
Tennessee Valley Authority Act	1933
Rural Electrification Act	1935
Food, Drug, and Cosmetic Act	1938
National Victory Garden Program	1941
National School Lunch Act	1946
Food for Peace Act	1954
Food Additives Amendment	1958
Food Stamp Act	1964
Child Nutrition Act	1966
Food Stamp Act	1970
National School Lunch Act – Amendments for Supplemental Nutrition for Women, Infants, and Children	1972
Agriculture and Consumer Protection Act	1973
FTC Improvement Act	1980
Nutrition Labeling and Education Act	1990
Food and Drug Administration Revitalization Act	1990
Children's Television Act	1990
Mickey Leland Childhood Hunger Relief Act	1994
Dietary Supplement Health and Education Act	1994
Personal Responsibility and Work Opportunities Reconciliation Act	1996
Food and Drug Administration Modernization Act	1997
Children's Online Privacy Protection Act	1998
Food Security and Rural Investment Act	2002
Food and Drug Administration Modernization Act	2007
Food, Conservation, and Energy Act	2008
American Recovery and ReinvestmentAct	2009
Healthy, Hunger-Free Kids Act	2010

Fig. 2.1 Important federal laws affecting food policy (sample)

The Congress passes laws that affect food policy. Figure 2.1 presents some of the federal laws over the past 75 years passed by the U.S. Congress that affect food policy.

Congress also acts by performing fact-finding that informs legislation. The principal government agency to carry out this work for Congress is the Congressional Research Service (CRS), which is a unit of the Library of Congress. It conducts legal and policy analyses as directed by Congress. For example, in 2010 the CRS did a data brief on childhood obesity for Congress (Corby-Edwards 2010). At other times, Congress wants to call upon the nation's scientific expertise. In these cases, Congress often commissions the National Institutes of Health, a private organization that is part of the National Academies of Science, to undertake a study for them. An example discussed in detail in Chap. 3 is a report evaluating 123 peer-reviewed scientific studies on the correlations between food marketing and children's food preferences, consumption, and health (Institute of Medicine 2005; Lewin et al. 2006).

The judicial branch both enforces the laws and sets regulatory frameworks through case law. In 1993, for example, 623 people in the western United States became ill and four children died from eating undercooked hamburger at Jack in the Box restaurants. Jack in the Box had ignored the warnings from both local health officials and their own employees that they were undercooking their hamburgers; the company did so because it believed that beef patties cooked to the recommended 155° came out too tough. The problematic hamburger they served was tainted with an unfamiliar strain of *E. coli* (O157:H7). In the 18 months following the incidents, the company lost $180 million and was faced with hundreds of lawsuits from individuals who became sick from consuming these hamburgers. In response, President Clinton called for Congressional hearings on food safety (Marler Clark n.d.).

To forestall government action, Jack in the Box adopted a food safety program known as Hazard Analysis and Critical Control Points, which had first been created by NASA in collaboration with the Pillsbury Company to reduce the risk of contaminated food for astronauts. The new program, which addressed practices at the slaughterhouse, beef in transit, and beef being prepared in the restaurant, greatly reduced illness from this strain of *E. coli*. The USDA Food Safety and Inspection Service Administrator designated the tainted hamburger as *adulterated* under the terms of the Federal Meat Inspection Act. This was the first time the term "adulterated" was used to refer to a microorganism that grew inside a cow; previously it had been applied only to harmful chemicals or foreign objects in food. After first unsuccessfully fighting the USDA in court, the beef industry (in particular, the National Cattleman's Beef Association and the American Meat Institute) spent more than $30 million on research to prevent these kinds of outbreaks. The industry tested various methods to kill these microorganisms and finally settled on the use of a technique known as "steam pasteurization," which had been developed by private industry but was certified in 1995 by the USDA as an approved method. It is a method commonly used, for example, by the large meat processors Tyson and Cargill. Large retailers such as Costco have also insisted on steam pasteurization or other methods that have been shown to be at least as effective (Andrews 2013). A story similar to the Jack in the Box story involves lawsuits related to poisoning from the consumption of raw Gulf oysters tainted with *Vibrio vulnificus* (Buzby and Frenzen 1999).

Even unsuccessful lawsuits can serve as a deterrent to harmful behavior. A good example is the Pelman class action suit against McDonalds for serving foods that lead to obesity. Following the suit, the corporation began to take steps—fitfully implemented—to post signs in its stores presenting nutritional information, remove *trans* fats and reduce saturated fats in its products, and provide more healthy alternative food choices such as apple slices (Mello et al. 2003; for more on lawsuits against food companies, see Meislik 2004).

The food industry can influence policy by using its deep pockets to lobby and advertise. Individuals and public interest nonprofits typically do not have the resources to match particular companies or industry trade organizations in these efforts. Instead, public interest groups use research and educational activities to inform the public, as well as lawsuits. Lawsuits—even if lodged against a particular company—can have the value of bringing public attention to a problem, motivate an entire industry to pay its fair share of costs, and redirect industry behavior—not just rectifying the actions of an individual company. Lawsuits concerning food policy are sometimes initiated by individuals, but more often they are initiated by public interest groups. For example, in 2006 a lawsuit was filed in the Massachusetts courts by the Center for Science in the Public Interest, Campaign for a Commercial-Free Childhood, and two Massachusetts parents against Viacom and Kellogg for marketing junk food to children (Center for Science in the Public Interest 2006; Nestle 2006).

State and local governments are also involved in setting food policy. For example, sugar-sweetened beverages in the schools are primarily regulated at the state level. Cigarettes are taxed at the state level, with wide variation in the amount of tax from state to state; it is most likely that it would be the states that would be the government body to leverage taxes on unhealthy foods. (See Chriqui et al. 2008 on state tax rates on snacks and sweetened beverages.) Local governments have also participated in food policy such as New York Mayor Michael Bloomberg's effort to ban the sale of sodas and other sweetened drinks in containers larger than 16 oz (later ruled to be unconstitutional) or San Francisco's "Happy Meal" law that bans free toys in meals targeted at children that do not meet a high nutritional standard. Local school boards often control the contracts with public schools about how the food industry can advertise and what can be served in vending machines on campus. Greves and Rivara (2006) offer a comparison of the competitive food policies in schools in 19 of the largest cities. Center for Science in the Public Interest (2007) gives a state-by-state review of food policies.

2.3 Informal Policy

Many parties play an informal role in establishing food policy. We have mentioned the role of individual companies and trade associations in lobbying the Executive and Legislative branches of federal, state, and local governments. Chapter 4 discusses former President Bill Clinton's role in using the bully pulpit to negotiate with the beverage industry to improve the healthiness of drinks available in public schools, while Chaps. 5 and 6 are focused on the use of the bully pulpit by First Ladies Eleanor Roosevelt and Michelle Obama. In a similar way, sports heroes and other celebrities can use their public familiarity as the bully pulpit for reform. For example, Beyoncé has supported Michelle Obama's initiative to fight childhood obesity; Ellen DeGeneres, Scarlett Johansson, and Jamie Oliver are all trading upon their fame to promote a healthy lifestyle (Conley 2011). Similarly, church authorities can take a position and use the consecrated pulpit as a bully pulpit for food policy. For example, Pastor Rick Warren, an influential California pastor who gave the prayer at the Obama inauguration in 2009, spoke against obesity, arguing for healthy eating and exercise using the "Daniel Plan," a plan for healthy eating and physical exercise named after the Biblical story of Daniel that was designed by three doctors including the television personality Mehmet Oz (Park 2012).

However, there are many other participants in the food policy debates. Professional organizations (as opposed to trade organizations) are active players.

For example, the American Academy of Pediatrics has written a policy paper on the impact of advertisements on children and adolescents (American Academy of Pediatrics 2006), while the Academy of Nutrition and Dietetics maintains web pages that track food and nutrition in public policy (eatright.org's Food and Nutrition in Public Policy page) and provides tips to the public about eating right (eatright.org's Public page).

Similarly, a number of private foundations and public interest groups have weighed in on the childhood obesity discussions. For example, the Robert Wood Johnson Foundation created the Healthy Kids, Healthy Communities program in 2008 to provide grants to local communities that are used to improve access to healthy foods in food deserts or improve the infrastructure for physical activity in their community (Ohri-Vachaspati et al. 2012; also see Levi et al. 2011 more generally about the Robert Wood Johnson Foundation).

A number of public interest groups are working on issues of food and children. The Rudd Center for Food Policy and Obesity convenes conferences on such topics as food and addiction and provides a clearinghouse for scientific research on food policy and obesity. The Center for Science in the Public Interest has been a watchdog for healthy food since it organized a campaign in 1973 to eliminate nitrates from bacon. The Alliance for a Healthier Generation, founded by the American Heart Association and the Clinton Foundation, brokers deals with companies and industries to improve the foods served in schools and also operates science-based programs in after-school environments (e.g., clubs and community centers) to improve healthy eating and physical activity in these places. The Alliance's work involves, for example, taking public stands on federal guidelines on snacks in schools. The Food Research and Action Center works with hundreds of organizations (nonprofits, labor unions, government agencies, and companies) to fight hunger in America. School Food FOCUS carries out policy work to improve the supply side of food so as to enhance food options in urban schools. These are among the most prominent public interest groups active in this policy sphere, but there are many others.

There is also a role to be played in food policy discussions by individuals. One common way for individuals to exert an influence is by using the media, such as writing a blog, preparing a YouTube video, or publishing a book. For example, Chap. 4 tells the story of Avis Richards, an independent filmmaker who produced a film criticizing the national school lunch program, and Mrs. Q (Sarah Wu), a schoolteacher who chronicled her experiences eating school lunches for a year in the Chicago public schools first through a blog, later in a book. (For a more detailed account of Mrs. Q, see Aspray et al. 2013.)

Individuals and groups of individuals sometimes use other legal means to capture a voice in the food policy debates. Tactics include boycotts, protests, petitions, letter-writing campaigns, strikes, work-to-rule, revelations, and teach-ins.

There are a number of instances of revelations, the making public or publicizing of facts about food and the food industry through writing or still or moving images. The cases of Avis Richards and Sarah Wu mentioned earlier are examples of revelations. A famous early example was Upton Sinclair's novel, *The Jungle*, published in 1906, which provided an exposé on life working in the slaughterhouses of Chicago. The public response to this book culminated in the passage of the Pure Food and Drug Act and the Meat Inspection Act that same year. A more recent example is journalist Eric Schlosser's nonfiction expose of the fast-food industry in America, *Fast Food Nation*, published in 2001. Undercover work by the Animal Liberation Front and People for the Ethical Treatment of Animals might also fit into this category, although their work has more often been about animal rights in research labs and product testing labs than about animals in food research and food production. These organizations send their members into labs undercover, posing as employees, to gain access to documents and while there they snap photographs with hidden cameras, which they release to the public in some form such as *Unnecessary Fuss*, a movie about the treatment of baboons in a University of Pennsylvania research laboratory.

There are examples of people boycotting food products. Perhaps the most famous was the grape boycott in the 1960s organized by the United Farmworkers Union to support the plight of the farmworkers who were picking grapes. More recently, there has been a campaign to boycott food products produced by Monsanto, Bayer CropScience, and other biotech companies that are made with genetically modified organisms (GMOs). For example, REALfarmacy.com provides a list of companies that use GMOs in their products, while one can find lists of non-GMO products on the web pages of The Non-GMO Project and The Institute for Responsible Technology (Huff 2013). A recent development is Buycott, an app for smart phones on the Android and iPhone platforms that enables an individual to scan the barcode on a product and determine whether the product should be boycotted (O'Connor 2013). Artist Sally Davies has taken periodic photographs of a Happy Meal she purchased from McDonalds that show the indestructibility of the product over time—up to day 1,125 at the time of this writing. One might consider this a cultural criticism or revelation of fast food as much as a boycott (Forbes 2010; Davies 2010).

There has also been use of petitions related to food products. For example, for a number of years the Coalition of Immokalee Workers had been seeking wage increases for tomato pickers in south Florida. In 1991 the Coalition changed its tactics to target the fast-food restaurants such as McDonalds and Taco Bell that served these tomatoes. In a campaign entitled Boot the Bell, the Coalition sent petitions and letters to Yum!, the parent corporation of Taco Bell, asking that it only purchase tomatoes from suppliers that paid the pickers at the higher pay rate. Boot the Bell expanded into the Campaign for Fair Food when various religious groups, labor unions, and student groups joined the effort. McDonalds and Yum! signed an agreement to pay the higher wages but Burger King refused. Later, under continued pressure, Burger King also settled (Gould-Wartofsky 2007; Hartford 2008).

Groups have also used protests against food companies and their practices. The Coalition of Immokalee Workers, mentioned earlier, protested for months in 2009 in front of Publix grocery stores because the chain continued to buy tomatoes from suppliers that did not pay the farmworkers the higher wages that McDonalds and Taco Bell were paying (Smith 2009). In 2011 a major gay rights organization, Human Rights Campaign, protested against Chick-fil-A for its support of an antigay marriage organization (Gilgoff 2011).

Individuals and groups of individuals can also participate in food policy discussions through civil disobedience. Examples include illegal boycotts, refusal to pay taxes, forbidden speech, threats to government officials, victimless crimes such as public nudity, riots, occupation of private property, denial of service attacks, information theft, and data leaks.

In 2013 there was a coordinated nationwide strike of fast food workers seeking a minimum hourly wage of $15 and the right to unionize. In St. Louis 12 of these protestors were arrested for failure to obey the reasonable commands of a police officer (KSDK 2013). Earlier the same month, in Seattle, under the organization Good Jobs Seattle, fast-food workers at Burger King, Taco Bell, Subway, Arby's, and Starbuck's, among other fast-food restaurants, picketed, carried out in-store demonstrations, offered a teach-in through the drive-through window, and finally eight workers linked arms and entered into civil disobedience outside a McDonalds restaurant, where they were arrested. They were protesting wage theft such as not paying at a higher rate for hours in excess of 40 per week, working without pay before or after their shifts, and taking illegal deductions from paychecks for such things as cash register shortages (Groves 2012).

Another type of illegal activity is vandalism. In 1992 Saeed Danosian, the manager of a McDonalds in Huntington Beach, California who had previously been trained as an artist in Vienna, painted a mural on the wall of a liquor store facing his McDonalds. The mural included representations of the Hamburglar, Ronald McDonald, and other McDonalds' characters. This mural had become a popular piece of public art in the city. One night in 2012 a radical group painted over the mural with the message "VEGAN" in large block letters. The original mural could not be repaired and had to be destroyed (Epting 2012; Arellano 2012).

We close this section with a description of the actions of two organizations, People for the Ethical Treatment of Animals (PETA) and the Animal Liberation Front (ALF). ALF is the more radical of the two organizations, although they often support one another's actions. Both have been designated by the USDA as terrorist threats (Frieden 2005; Merchant 2009). There has been a pattern of violence, arson, and assaults that have been attributed to these organizations.

In 2000 a PETA activist threw a pie in the face of the USDA Secretary at the National Nutrition Summit (Southern Policy Law Center 2002). Playing off the cardboard crowns that Burger King traditionally provided to children as a promotional item, in 2001 PETA began distributing crowns soaked in blood outside select Burger Kings across the United States and Canada as a means to protest the treatment the animals receive that end up in Burger King meals (Johnson and Johnson 2001). That same year, actor James Cromwell and three PETA officials were arrested at a Wendy's restaurant in Fairfax, Virginia for refusing to leave the premises. They were there to protest the company's treatment of pigs and chickens. Cromwell told the press, "after *Babe*, people recognized that pigs and other animals abused on factory farms are sensitive, gentle animals. It is high time that big corporations like Wendy's stopped treating these wonderful animals like meat machines" (Sims 2001). In 2005 PETA organized a campaign targeted at making children vegetarians, by depicting parents as "hooked on killing" and advising children to keep puppies and kittens away from dad if he spends leisure time fishing (FoxNews 2005). This followed a cartoon style pamphlet distributed by PETA in 2003 entitled "Your Mommy Kills Animals" (PETA Kills Animals 2013). In 2009 PETA released a game Super Chick Sisters, a parody of Super Mario Brothers that details McDonald's mistreatment of animals (Fahey 2009). The following year, PETA released another new game, Super Tofu Boy, which is a parody of Super Meat Boy (Fahey 2010).

Some of ALF's activities are more radical than those of PETA. For example, in 1987 "ALF" and "murderers" are the words that were painted on the bailding at a fire causing $200,000 in damage to the V. Melani poultry distribution company. In 1989, ALF burned down an Egg Products store in Salt Lake City and destroyed two of the company trucks. In 1997 it sprayed a noxious chemical in a McDonalds restaurant in Troy, Michigan and spray painted "McShit, McMurder, McDeath" on the bathroom walls. That same year, ALF burned down a McDonalds restaurant in West Jordan, Utah. In 1999 ALF set a fire in the Childer's Meat Co. in Eugene, Oregon, causing extensive damage (Southern Policy Law Center 2002).

ALF continues to be active. In 2007 it took credit for the assault on a KFC fast-food restaurant in Bremerton, Washington, where "Animal Love," "Mess with Animals get Served," and "Meat is Murder" were spray painted on the side of the building and "Boycott KFC" posters were plastered to the exterior (Kitsap Sun Staff 2007). In 2011 ALF took credit for tearing down the fencing surrounding a pen at Damascus Elk Farm in Clackamas, Oregon (Animal Liberation Front 2011). In 2012, 75–100 pheasants were released from a farm in Scio, Oregon that breeds ring-necked pheasants for hunting and dog training. In an anonymous post on the Bite Back magazine website, The Animal Liberation Front took credit for dismantling a pheasant aviary and liberating the animals into the night sky (KVAL News Staff 2012). In 2013, ALF took credit for inserting glue in locks, pouring red paint, and spray painting the words "Free the Animals" across the patio of the Taco Asylum Restaurant in Costa Mesa, CA. ALF claimed to have targeted Taco Asylum for selling the meat of rabbits, ducks, cows, and pigs (Schou 2013).

This chapter has surveyed a number of ways in which formal and informal approaches to food policies are carried out. The next two chapters present detailed discussions of two important aspects of the food-related policy issue of childhood obesity. The final two chapters look not at a single food policy issue, such as childhood obesity, but instead give two examples of one particular type of informal policy approach, the use of the bully pulpit by First Ladies, as these two women addressed myriad food policy issues.

References

Alderman J, Smith JA, Fried EJ, Daynard RA (2007) Application of law to the childhood obesity epidemic. J Law Med Ethics 35(1):90–112

American Academy of Pediatrics (2006) Children, adolescents, and advertising. Policy Statement, Committee on Communications. Pediatr 118(6):2563–2569

Andrews J (2013) Jack in the Box and the decline of E. coli. Food Safety News. http://www.foodsafetynews.com/2013/02/jack-in-the-box-and-the-decline-of-e-coli/#.Uic-SaVxm4h, 11 Feb 2013, Accessed 4 Sept 2013

Anomaly J (2012) Is obesity a public health problem? Public Health Ethics 5(3):216–221

Arellano G (2012) (November 13). "Vegans vandalize beloved Huntington Beach mural with McDonald's characters on it". OC Weekly. http://blogs.ocweekly.com/stickaforkinit/2012/11/saeed_danosian_mcdonalds.php. 13 Nov 2012, Accessed 6 Sept 2013.

Aspray W, Royer G, Ocepek MG (2013) Food in the Internet Age. Springer, New York

Buzby JC, Frenzen PD (1999) Food safety and product liability. Food Policy 24:637–651

Center for Science in the Public Interest (2006) Parents and advocates will sue Viacom & Kellogg. http://www.cspinet.org/new/200601181_print.html. 18 Jan 2006, Accessed 5 Sept 2013

Center for Science in the Public Interest (2007) State school foods report card: A state-by-state evaluation of policies for foods and beverages sold through vending machines, school stores, a la carte, and other venues outside of school meals. Center for Science in the Public Interest, Washington DC. http://www.cspinet.org/2007schoolreport.pdf. Accessed 16 Dec 2013

Chriqui JF, Eidson SS, Bates H, Kowalczyk S, Chaploupka FJ (2008) State sales tax rates for soft drinks and snacks sold through grocery stores and vending machines. J Public Health Policy 29:226–249

Conley M (2011) Beyonce joins Michelle Obama's initiative to fight childhood obesity. ABC News. http://abcnews.go.com/Health/beyonce-drops-music-video-fight-childhood-obesity/story?id=13482133. 29 April 2011, Accessed 3 Sept 2013

Corby-Edwards AK (2010) (Sept 21). Childhood overweight and obesity: Data brief. Congressional Research Service R41420. http://congressional.proquest.com/profiles/gis/result/pqpresultpage.gispdfhitspanel.pdflink/http:$2f$2fprod.cosmos.dc4.bowker-dmz.com$2fapp-bin$2fgis-congresearch$2f7$2fc$2f6$2f7$2fcrs-2010-fdt-0792_from_1_to_20.pdf/entitlementkeys=1234%7Capp-gis%7Ccongresearch%7Ccrs-2010-fdt-0792. 21 Sept 2010, Accessed 18 Nov 2013

Davies S (2010) Davies McDonalds Happy Meal Project www.flicker.com/photos/sallydavies/sets/72157624739645253/ Accessed 26 Feb 2014

Epting C (2012) In the pipeline: Vegan vandalism not kosher. Huntington Beach Independent. http://www.hbindependent.com/opinion/tn-hbi-1115-in-the-pipeline-20121112,0,2313248.story. 12 Nov 2012, Accessed 6 Sept 2013

Fahey M (2009) PETA releases new Super Chick Sisters. Kotaku. http://kotaku.com/5403312/peta-releases-new-super-chick-sisters. 12 Nov 2009, Accessed 13 Sept 2013

Fahey M (2010) PETA roasts Super Meat Boy with tofu. Kotaku. http://kotaku.com/5703319/peta-roasts-super-meat-boy-with-tofu. 1 Dec 2010, Accessed 13 Sept 2013

Forbes P (2010) Six-month-old McDonald's Happy Meal is indestructible. National Pages, Eater. http://eater.com/archives/2010/10/13/sixmonthold-happy-meal-is-indestructible.php. 13 Oct 2010, Accessed 6 Sept 2013

FoxNews 2005 PETA tells kids to run from daddy. Fox News. http://www.foxnews.com/story/2005/11/25/peta-tells-kids-to-run-from-daddy/. 25 Nov 2005, Accessed 13 Sept 2013

Frieden T (2005) FBI, ATF address domestic terrorism. CNN. http://www.cnn.com/2005/US/05/19/domestic.terrorism/. 19 May 2005, Accessed 6 Sept 2013

Frieden TR, Dietz W, Collins J (2010) Reducing childhood obesity through policy change: Acting now to prevent obesity. Health Aff 29(3):357–363

Gilgoff D (2011) Chick-fil-A controversy shines light on restaurant's Christian DNA. CNN Belief Blog. http://religion.blogs.cnn.com/2011/02/04/chick-fil-a-controversy-shines-light-on-restaurants-christian-dna/?hpt=C2. 4 Feb 2011, Accessed 6 Sept 2013

Gould-Wartofsky M (2007) Farmworkers and students take on Burger King. The Nation. http://www.thenation.com/article/farmworkers-and-students-take-burger-king#axzz2e7wcQbUZ. 14 Dec 2007, Accessed 6 Sept 2013

Greves HM, Rivara FP (2006) Report card on school snack food policies among the United States' largest school districts in 2004–2005: Room for improvement. Int J Behav Nutr Phys Act 3:1. doi:10.1186/1479-5868-3-1

Groves D (2012) Arrests at protest of fast-food wage theft in Seattle. The Stand. http://www.thestand.org/2013/08/8-arrested-protesting-fast-food-wage-theft/. 2 Aug 2012, Accessed 6 Sept 2013
Hartford J (2008) Tomato Troubles. QSR Magazine. http://www2.qsrmagazine.com/articles/exclusives/0208/immokalee-1.phtml., Feb 2008, Accessed 6 Sept 2013
Huff EA (2013) Boycott these Monsanto-pandering factory 'food' companies and help drive GMOs out of the food supply. Natural News. http://www.naturalnews.com/040758_GMO_boycott_Monsanto_food_companies.html. 13 June 2013, Accessed 6 Sept 2013
Institute of Medicine (2005) Food marketing to children and youth: Threat or opportunity? National Academies Press, Washington DC
Ippolito PM (1999) How government policies shape the food and nutrition information environment. Food Policy 24:295–306
Johnson H, Johnson H (2001) Animal rights groups protest fast food chain. Indiana Daily Student. http://www.idsnews.com/news/story.aspx?id=17250. 18 June 2001, Accessed 6 Sept 2013
Khan F (2011) Combatting Obesity through the Built Environment: Is There a Clear Path to Success? J Law Med Ethics 39(3):387–393
Kitsap Sun Staff (2007) Vandals tag fast-food restaurant. http://www.kitsapsun.com/news/2007/dec/14/fast-food-restaurant-allegedly-tagged-by/#axzz2ehZ7rrTQ. 14 Dec 2007, Accessed 13 Sept 2013
KSDK (2013) 12 arrests made at St. Louis fast food protests. http://www.ksdk.com/news/article/395480/3/12-arrests-made-at-St-Louis-fast-food-protests. 29 Aug 2013, Accessed 6 Sept 2013
KVAL News (2012) Sheriff: Freeing pheasants from farm 'a form of domestic terrorism.' http://www.kval.com/news/business/ALF-pheasants-Scio-143370956.html. 19 March 2012, Accessed13 Sept 2013
Leibowitz J (2008) Concurring statement of Commissioner Jon Leibowitz, Marketing food to children and adolescents: A review of industry expenditures, activities, and self-regulation. http://www.ftc.gov/public-statements/2008/07/concurring-statement-commissioner-jon-leibowitz-regarding-marketing-food. 29 July 2008, Accessed 16 Dec 2013
Levi J, Siegel L, St. Laurent R, Kohn D (2011) "F as in fat: How obesity threatens America's future". Trust for America's Health and the Robert Wood Johnson Foundation. http://www.rwjf.org/content/rwjf/en/research-publications/find-rwjf-research/2011/07/f-as-in-fat.html. Accessed 16 Dec 2013
Lewin A, Lindstrom L, Nestle M (2006) Food industry promises to address childhood obesity: Preliminary investigation. J Public Health Policy 27:327–348
Marler Clark Attorneys at Law (nd) Jack in the Box E. coli outbreak lawsuits. http://www.marlerclark.com/case_news/view/jack-in-the-box-e-coli-outbreak-western-states. Accessed 5 Sept 2013
McGinnis J, Meyers M, Meyers LD (1999) Public policy and healthy eating. Food Policy 24(2–3):335–341
Meislik A (2004) Weighing in on the scales of justice: The obesity epidemic and litigation against the food industry. Ariz Law Rev 46:781–788
Mello MM (2010) Federal Trade Commission regulation of food advertising to children: Possibilities for a reinvigorated role. J Health Polit Policy Law 35(2): 227–276
Mello MM, Rimm EB, Studdert DM (2003) The McLawsuit: The fast-food industry and legal accountability for obesity. Health Aff 22(6):207–216
Mello MM, Studdert DM, Brennan TA (2006) Obesity: The new frontier of public health law. N Engl J Med 354(24):2601–2610
Merchant B (2009) USDA classifies PETA as a terrorist threat. Treehugger. http://www.treehugger.com/corporate-responsibility/usda-classifies-peta-as-a-terrorist-threat.html?campaign=th_rss&utm_source=feedburner&utm_medium=feed&utm_campaign=Feed%3A+treehuggersite+(Treehugger)&utm_content=Google+Reader. 2 Dec 2009, Accessed 6 Sept 2013
Moorman C, Price LL (1989) Consumer policy remedies and consumer segment interactions. J Public Policy Mark 8:181–203

National Agricultural Hall of Fame (nd) Henry Leavitt Ellsworth. http://www.aghalloffame.com/hall/ellsworth.aspx. Accessed 5 Sept 2013

Nestle M (2006) Food marketing and childhood obesity: A matter of policy. N Engl J Med 354(24):2527–2529

O'Connor C (2013) New app let's you boycott Koch brothers, Monsanto and more by scanning your shopping cart. Forbes. http://www.forbes.com/sites/clareoconnor/2013/05/14/new-app-lets-you-boycott-koch-brothers-monsanto-and-more-by-scanning-your-shopping-cart/. 14 May 2013, Accessed 6 Sept 2013

Ohri-Vachaspati P, Leviton L, Bors P, Brennan L, Brownson RC, Strunk S (2012) Strategies proposed by healthy kids, healthy communities partnerships to prevent childhood obesity. Prev Chronic Dis 9:100292. doi:10.5888/pcd9.100292

Park M (2012) Rich Warren and church tackle obesity. CNN http://www.cnn.com/2012/01/24/health/saddleback-warren-diet. 24 Jan 2012, Accessed 3 Sept 2013

Perdue WC, Gostin LO, Stone LA (2003) Public health and the built environment: Historical, empirical, and theoretical foundations for an expanded role. J Law Med Ethics 31:557–566

PETA Kills Animals (2013) PETA targets your kids. http://www.petakillsanimals.com/children. Accessed 18 Nov 2013

Schou N (2013) Animal Liberation Front vandalizes Taco Asylum restaurant. OC Weekly. http://blogs.ocweekly.com/navelgazing/2013/08/animal_liberation_front_vandal.php. 8 Aug 2013, Accessed 13 Sept 2013

Schwartz MB, Brownell, KD (2007) Actions necessary to prevent childhood obesity: Creating the climate for change. J Law Med Ethics 35(1):78–89

Seiders K, Petty RD (2004) Obesity and the role of food marketing: A policy analysis of issues and remedies. J Public Policy Mark 23(2):153–169

Sims R (2001) Cromwell arrested during protest. Hollywood. http://www.hollywood.com/news/brief/470673/cromwell-arrested-during-protest?page=all. 4 July 2001, Accessed 6 Sept 2013

Smith C (2009) Farmworkers protest supermarket tomatoes. The St. Augustine Record. http://staugustine.com/news/local-news/2009-11-24/farmworkers-protest-supermarket-tomatoes#.UioQ1qVxm4h. 24 Nov 2009, Accessed 6 Sept 2013

Southern Policy Law Center (2002) Eco-Violence: The record. Intell Rep107. http://www.splcenter.org/get-informed/intelligence-report/browse-all-issues/2002/fall/from-push-to-shove/eco-violence-the-rec, Accessed 6 Sept 2013

Sugarman SD, Sandman N (2007) Fighting childhood obesity through performance-based regulation of the food industry. Duke Law J 56:1403–1490

Thaler RH, Sunstein C (2008) Nudge: Improving decisions about health, wealth, and happiness. Yale University Press, New Haven

Waggoner PE (1976) Research and education in American agriculture. Agricultural History. 50(1):230–247

Young P (2011) Animal Liberation Front (A.L.F.) liberates elk from Oregon farm. Animal Liberation Frontline. http://www.animalliberationfrontline.com/animal-liberation-front-a-l-f-liberates-elk-from-oregon-farm/. 4 Sept 2011, Accessed 13 Sept 2013

Chapter 3
Protecting Children from Obesity: A History of Television and Internet Food Advertising Regulation in the United States

It's just hard not to listen to TV: it's spent so much more time raising us than you have.

(Bart Simpson, Quotes on the Media and Children n.d.)

And somebody asked me: 'Lucy, is that ethical? You're essentially manipulating these children.' Well, is it ethical? I don't know. But our role at Initiative is to move products, and if we know you move products with a certain creative execution, placed in a certain type of media vehicle, then we've done our job. They are tomorrow's consumer—tomorrow's adult consumer—so start talking with them now, build that relationship when they're younger, and you've got them as an adult.

(Lucy Hughes [from The Corporation], IMDB.com 2013)

Companies have moved away from exaggerating the product characteristics to a whole new form of advertising, which is symbolic advertising. The product is pushed not on the basis of what it can do, or how it tastes, but of its social meaning. So kids are taught to want candy, or sugared cereals, or soda because it's cool. It will define them as an individual. What you buy is who you are.

(Juliet Schor, IMDB.com 2013)

This chapter examines one important aspect of the national problem of obesity among American children and the attempts to resolve it by political means. While there are many root causes to the epidemic in childhood obesity, this chapter focuses on only one of them: advertising by food companies to children in their homes, which many researchers believe entices children to eat the wrong kinds of foods in excessive quantities. "Enticement" is a strong word, but it is applicable in this case. Low-nutrient food advertising uses images of coolness and fun, free games, well-known characters from television and movies, prizes, and other means to influence

W. Aspray et al., *Formal and Informal Approaches to Food Policy*, SpringerBriefs in Food, Health, and Nutrition, DOI 10.1007/978-3-319-04966-3_3,

Relation between eating fast-food and obesity	Spencer, Frank, and McIntosh 2005; Rosenheck 2008; Chou, Rashad, and Grossman 2008; Malik, Schultze, and Hu 2006; Pereira et al. 2005
Advertising rather the sedentary practice of watching television that mainly contributes to obesity	Zimmerman and Bell 2010
Insufficient cognitive maturity to evaluate an advertiser's claims or possibly not recognize the difference between advertising and other programming	Harris, Bargh, Brownell 2009; Connor 2006; John 1999; Chernin 2008; Livingstone and Helsper 2006
Constitutional rights of children as a vulnerable population	Harris et al. 2009

Fig. 3.1 Related scientific literature on children, television watching, food, and obesity

a young population, some of whom do not have the cognitive maturity to evaluate an advertiser's claims or possibly not even recognize the difference between advertising and other programming. Thus, the overall story presented here is one about a vulnerable population (children) and the efforts by adults to protect this population from these enticements by political means. Figure 3.1 provides some of the background scientific literature on children, television watching, food, and obesity that underpins this discussion. (For an alternate and decided by minority approach to the causes of childhood obesity, see Cornwell and McAlister (2011).)

The concern about the harmful effects of food advertising on children's health is similar in many respects to concerns about gambling, violent video games, alcoholic drinking, and especially cigarette smoking among children; it is a part of a larger narrative concerning the politics of enticement in America. Figure 3.2 presents examples of the research literature on advertising these other enticements to children. The literature on cigarettes is by far the largest, and there are many parallels between the advertisement to children of nonnutritious foods and the advertisement of cigarettes.

The analysis in this chapter is presented in five parts, following a roughly chronological account. The first section discusses efforts in the 1970s in which U.S. federal agencies attempted with limited success to regulate food advertising to children on television—first proposed to resolve an epidemic of tooth decay rather than an epidemic of obesity. The second section focuses on the following quarter century, during which the political climate had generally moved away from federal regulation. Practically the only regulation of food advertising during this era was self-regulation by industry. This period was characterized by weak regulatory efforts and growing problems with childhood obesity. The third section discusses the efforts during the first decade of the twenty-first century to find new ways to fight childhood obesity as it became increasingly apparent to the scientific community and the general public that the nation faced a serious problem. The final two sections examine two major risks to any of the proposed solutions to regulation of food advertising—one section discusses the argument by industry that companies have a First Amendment right to free commercial speech based on court decisions appearing between 1976 and 2001; the other section discusses the issues that arose as food advertising expanded into the new medium of the Internet. A short final section offers some conclusions.

Cigarettes	Schor and Ford 2007; Courtney 2006; Frieden, Dietz, and Collins 2010; Brownell and Horgen 2004; Montgomery and Chester 2009; Sargent, Gibson, and Heatherton 2009; Kline et al. 2006; Nelson 2006; Hanewinkel et al. 2010; Morrison, Krugman, and Park 2008; Krugman and King 2000; Bayer and Kelly 2010; Celebucki and Diskin 2002; Luke et al. 2011; DiFranza et al. 1991; Shadel, Tharp-Taylor, and Fryer 2009; Hawkins and Hane 2000; Charlesworth and Glantz 2005; Sebrie and Glantz 2007; Forsythe and Malone 2010; Gostin 2009; Capella, Taylor, and Webster 2008; Sung and Hennink-Kaminski 2008; King et al. 1998; Blum 2010; Pierce et al. 2010; Givel 2007; Hoek et al. 2010; King and Siegel 2001; Goldstein et al. 1987; Henriksen 2010; Ciolli 2007; McCool et al. 2012; Freeman et al. 2009; DiRocco and Shadel 2007; Kelly, King, et al. 2011;Carter, Mills, and Donovan 2008; Pollay et al 1996; Henrikson et al. 2008; and Botvin et al. 1993
Alcohol	Goldfarb and Tucker 2010; Hebden 2011; Anderson et al. 2009; Nelson 2010; Gentry et al. 2011; Gunter, Hansen, and Touri 2009; and Barry and Goodson 2010
Violent video games	Barlett and Anderson 2007; Bijvank et al. 2009; Rose-Steinberg 2010; Chang 2010; Hunter, Lozada, and Mayo 2011; Strasburger 2009; Wojciechowski 2010; Kenyota 2008; Becker-Olsen and Norberg 2010; and Collier, Liddell, and Liddell 2008
Gambling	Monaghan, Derevensky, andSklar2008

Fig. 3.2 Research literature on advertising other enticements to children

An important theme running throughout this chapter is that the efforts to regulate television advertising have been shaped by the general attitudinal climate toward federal regulation of industry among the American public. At times, the public has believed that industry is too powerful and acts in its own self-interest in ways that must be reined in by federal regulation. At other times, there has been greater public trust in industry and a belief that regulation inhibits competition and decreases economic efficiency. In these times of antiregulatory sentiment, industry has pushed for the replacement of government regulation with industry self-regulation and voluntary compliance.

The public move toward government regulation of industry has been cyclical. The greatest period of regulation was during the Progressive Era in the early twentieth century. Proregulatory sentiment died off during the 1920s but returned as part of the New Deal legislation to overcome the social effects of the Great Depression. Concern about abuse of government regulatory power led to new limitations on the policing powers of federal agencies after the Second World War. However, support for federal regulation picked up once again during President Johnson's Great Society program in the 1960s, but by the 1970s there was again a move toward less regulation of industry. In the 1980s, the Reagan administration attempted substantive deregulation of industry and placed new restrictions on the right of agencies to regulate. There has been a long run of popular support for deregulation since the 1980s. The exceptions during this era have been made selectively—to return to regulation of specific areas precipitated by scandals such as the savings and loan crisis and failure of Enron while leaving deregulation in place in other spheres. Even today, the U.S. population still largely favors deregulation. (Cyclical American attitudes toward industry regulation and its implications for childhood obesity were noted by Alderman et al. (2007), who points the reader to a strong literature on the history of industrial regulation in America: Croley 2003; Estlund 2005; Glaeser and Schliefer 2003; Hanson and Yosifon 2003; Kahn 2002; Rabin 1986; Rose-Ackerman 1990; Rubin 2005.)

The first efforts to regulate television advertising of food to children occurred in the 1970s, led by government bureaucrats who had been installed in their jobs when federal regulation of industry was ascendant. However, just at the time the Federal Communications Commission (FCC) and the Federal Trade Commission (FTC) began to consider regulation of advertisements of nonnutritious foods on children's television, public sentiment was turning against government regulation. Antiregulatory sentiment has continued ever since, and the only thing that has enabled a stronger government hand in controlling industry has been the overwhelming evidence that there is a worsening epidemic of childhood obesity in America, which became apparent around 2000. The large number of cases of childhood obesity makes this an exceptional case, much like the savings and loan crisis.

The overarching theme of this book is the interplay between formal and informal approaches to food policy. The complexity of this interaction can be appreciated by simply listing the major players in the policy area covered by this chapter, viz., the regulation of food advertising directed at children appearing on television or the Internet. On the formal policy side, there are the federal agencies (FCC, FTC, Department of Health and Human Services) and Congress (which passed various relevant pieces of legislation including the Children's Television Act, FTC Improvement Act, Children's Online Privacy Protection Act (COPPA), American Recovery and Reinvestment Act). The Institute of Medicine has an interesting position. It has such a positive reputation in Washington that it is often treated as part of the formal policy-making process even though it and the other arms of the National Academies of Science are actually private, nonprofit organizations. On the informal policy side in support of less federal regulation, there are media companies (Disney), media trade associations (National Association of Broadcasters), food companies (McDonald's, Kraft Foods), advertising trade associations (National Advertising Review Council), general business trade associations (Council of Better Business Bureaus (CBBB)), and industry associations put together for a specific purpose in this food policy battle (Children's Advertising Review Unit, Children's Food and Beverage Initiative, Sensible Food Policy Coalition). On the proregulatory side of these policy battles are private foundations (Robert Wood Johnson Foundation), professional associations (American Psychological Association, American Academy of Pediatrics), and public interest nonprofits (Action for Children's Television; Center for Science in the Public Interest Consumers Union; Committee on Children's Television; Council on Children, Media, and Merchandising).

It is beyond the scope of this chapter to try to differentiate political action among industries (e.g., broadcast versus food) and among players in a given industry (e.g., difference among individual firms within the food industry and between the firms and their industry trade associations). While occasionally an individual company will act out of step with the rest of its industry—and we mention a few instances—there is a great deal of uniformity within the corporate sector involved with advertising food to children on television and the Internet. (For more on corporate action, see Hillman et al. 2004; Scholzman 2011; Schuler et al. 2002.)

3.1 Television Advertising of Food to Children: Early Efforts at Regulation by the FCC and the FTC

Much of the public concern and policy response to childhood obesity has surrounded the advertising to children of low-nutrient foods high in fat and sugar, especially sugared drinks, sweetened cereals, and fast food meals. Advertising on children's television programming has been identified since the 1970s as a major factor contributing to the high frequency with which these low-nutrient foods appear in the diets of American children.

When the public first began to become concerned with television food advertising to children, there was only one major television show directed at preschoolers, *Captain Kangaroo*. It showed on network television from 1955 to 1984. However, several major changes occurred to make children's television ads more prevalent and hence of greater concern. During the decade of the 1960s, the national television networks regularized their programming to show cartoons every Saturday from 8:00 a.m. until noon. This concentration of children's programming was designed mainly to attract advertisers. With the growth of cable television in the United States in the 1970s came the rise of cable networks directed at children. Nickelodeon, providing daytime children's programming, was founded in 1979. In 1985 Nickelodeon created Nick at Night, offering in the evenings and overnight family programming (also of interest to children) such as reruns of *Bewitched* and the *Mary Tyler Moore Show*. Competition appeared on cable and satellite television, such as the Disney Channel (1983) and the Cartoon Network (1992). (For a discussion of branding on children's television networks and its influence on children, see Preston and White 2004; Connor 2006; Robinson et al. 2007.)

The amount of television watched by children grew steadily over the final three decades of the twentieth century. The number of television sets in American households grew, and increasingly television receivers were placed in children's bedrooms, where they could be watched without parental supervision. By the year 2000, children spent more time (1,250 h per year on average) watching television than they spent in school—in fact more hours than they did doing any activity other than sleeping (Byrd-Bredbenner 2002; Byrd-Bredbenner and Grasso 2000; Holt et al. 2007; Desrochers and Holt 2007).

During the last three decades of the twentieth century, the frequency of food television advertising also increased. This is not surprising, given that the food industry is the second largest advertiser in the United States, after the automobile industry (Story and French 2004). One-eighth of all consumer dollars are spent on food purchases; given that food is a repeat purchase and highly branded, there is great incentive for the food industry to advertise (Story and French 2004). It is believed today that middle- and upper-class white American children see between 20,000 (Schor and Ford 2007) and 40,000 (Kunkel et al. 2004; also see Powell et al. 2007; Holt et al. 2007) ads each year, and that poor, black, and Hispanic children have even greater exposure (Grier and Kumanyika 2008).

Food advertising directed at children goes back at least as far as the 1930s, when the Mickey Mouse character was licensed for use by Post Toasties cereal. Beginning in the 1960s, and continuing throughout the rest of the century, ads for foods typically consumed by children were increasingly targeted at the children themselves rather than at their parents. More than 80 % of children's ads were for toys, cereals, candies, and fast-food restaurants (Byrd-Bredbenner 2002; also see Batada et al. 2008; Folta et al. 2006; Powell et al. 2007). The vast majority of the food ads were for low nutrition, highly sugared foods (Ippolito and Pappalardo 2002). Breakfast cereal ads were particularly common in the 1970s, but the number of fast-food restaurant ads grew, especially during the 1990s. It was hard to watch a Saturday morning of programming without seeing ads for Kellogg's Frosted Flakes, McDonald's Happy Meals, or Skittles candy. (For an interesting national comparison, see studies examining food advertising directed at children in Canada, including Nadeau 2011; Dhar and Baylis 2011; Richards and Padilla 2009; Kent et al. 2012.)

The general consensus among the public health advocates, if not among the advertising and food industry executives, was that food marketing was having a significant negative impact on children's food consumption, taste, and health. The ads, it was believed, were leading to "hedonic hunger urges" in the absence of energy deficits, encouraging snacking at nonmeal times, consuming less healthy food choices, establishing more materialistic values in children, and contributing to health problems such as obesity and high cholesterol. There was particular concern about the impact on the relationship between children and their parents. Three quarters of parent–child communications about products involve children demanding things they have seen on television. Children begin to nag for products as early as age 24 months, and 75 % of the time the first instance of nagging occurs in a supermarket. Parents frequently yield to their children's "pester power" and, when they do not yield, the children often become angry or disappointed. (See, for example, Hastings et al. 2003; Story and French 2004; Lowe and Butryn 2007; Galst and White 1976; Nadeau 2011; Pettersson et al. 2004; Center for Science in the Public Interest 2003 on these points. For a theoretical framing of the persuasive power of advertisement upon children, see Buijzen et al. 2010.)

Advertisers were targeting children of almost every age with their ads. The ads on television programs that targeted preschoolers focused mainly on building brand recognition and loyalty, not on immediately selling products. They used images of children or licensed figures (e.g., Ronald McDonald or Tony the Tiger) having fun or doing exciting things; sometimes there were few or no food images in the ads (Connor 2006). For older children and adolescents, there were efforts not only to build brand recognition and loyalty, but also to market specific products. Marketers' hope was that children would either pester their parents to buy these products or, as teenagers with disposable income, consider buying them for themselves. One study identified the strategies and elements in television commercials from 2004 to 2005 targeting 8- to 12-year-olds: "The most frequent promotional strategies were the use of jingles/slogan, showing children with food, use of product identification characters, cross-selling of toys, and being directed to a website. The most common attention elements were showing 'real children', animation, animals with human characteristics, fast-cutting scenes, exciting/fast-paced music, humor, and color effects" (Page and Brewster 2007).

In 1968, when Peggy Charren became concerned with the lack of quality television programming for her 4-year-old daughter, she formed the nonprofit group Action for Children's Television (ACT) (Lawson 1991; on the general history of children and television, see Pecora et al. 2006; Huston et al. 1990; Leifer et al. 1974; Zimmerman and Christakis 2005; Hofferth and Sandberg 2001). Over the next several years, ACT worked with other large national organizations including the American Academy of Pediatrics, the National Education Association, the National Parent Teachers Association, and the Center for Science in the Public Interest (CSPI) and its spinoff organization, the Center for Study of Commercialism, as well as many religious organizations. Although critics referred derogatively to ACT as a bunch of housewives, it was operated in a highly professional manner and grew by the late 1970s to employ a staff of 15 (dropping to a quarter of that size a decade later and disbanding in 1992). ACT pursued its battles mainly in the courts (Lawson 1991; New York Times 1992; Pecora 2007).

Responding in part to calls from ACT to either limit or eliminate ads on children's television programming, both the FTC, which regulates advertising, and the FCC, which regulates television, held hearings (Story and French 2004; Alderman et al. 2007; Uscinski 1984). The FCC took the first action—in 1974—calling not only for limits on the number of minutes of advertising on children's television programming, but also requiring the introduction of rhetorical devices that more clearly separated programming from advertising ("...and now a word from our sponsor") and banning host selling (the process by which characters from the show are used to advertise goods in the commercials that air during that show). Interestingly, ACT was concerned not so much about the number of junk food ads but instead about the large number of vitamin ads directed at children. (For a study of host selling and the associated legal issues, see Campbell 2006.)

There were already concerns in the scholarly community about the effect of advertising on children. Later research provided evidence that advertising to children contributes to materialism, life unhappiness, parent–child conflict, disappointment, and dissatisfaction (Zuckerman and Zuckerman 1985; Buizjen and Valkenburg 2003a, b). Regulatory efforts targeting children's advertising had previously been used for public health purposes, notably including the Public Health and Cigarette Smoking Act of 1970, which banned cigarette advertising on radio and television as of 1971 (Hamilton 1972; Holak and Reddy 1986; Eckard 1991; Pollay 1995). Thus, there was a wide-held belief in regulation as an effective means of serving public health goals.

In 1973, before the FCC could take action, the National Association of Broadcasters (NAB) took measures to avoid federal regulation by voluntarily changing its Television Code so as to reduce the amount of time allotted to ads during children's programming. After several early tweaks, the Association settled on an advertising limit of 9.5 min per hour on weekends and 12 min per hour on weekdays. (The NAB is the trade organization for the U.S. commercial broadcast industry—formed in 1922 as the National Association of Radio Broadcasters, renamed the National Association of Radio and Television in 1951 and the NAB in 1958. It lobbies on various policy issues on behalf of the broadcast industry and carries out various educational and research activities concerning First Amendment rights through its associated foundation.) As part of its licensing renewal standards, in 1974 the FCC adopted rules for time allotted to ads in children's programming that mirrored those set by industry (Kunkel and Watkins 1987; Kunkel 1991; Campbell 1999).

In 1984, in a deregulatory move originating in the Reagan administration, the FCC rescinded all limits on advertising during children's programming, leading to a rapid increase in the number of ads appearing (Byrd-Bredbenner 2002). The Children's Television Act was passed overwhelmingly by Congress in 1988 to address this problem, but the Act was vetoed by President Reagan, who called it an assault on freedom of expression. This led to another wave of increase in the number of ads shown on children's programming. Finally, in 1990, The Children's Television Act was enacted as law. The voluntary restrictions from the broadcasting industry remained in effect all of this time, until the late 1990s, when the Justice Department ruled that the voluntary Television Code was a violation of antitrust law and its time restrictions were thereby eliminated (Byrd-Bredbenner 2002; Jordan 2008; Mello 2010).

The Children's Television Act did not regulate the content of the advertisements, only the amount of advertising that appeared on children's programming (setting limits of 10.5 min per hour on weekends and 12 min per hour on weekdays, which were again embodied in the FCC broadcast license renewal standards). During 1977 and 1978, four public interest groups (ACT, the CSPI, Consumers Union, and the Committee on Children's Television) filed petitions with the FTC to take action against food advertisements directed at children. According to the authority granted to them by Congress, the FTC could control both unfair and deceptive ads, with the right to both make regulations and bring lawsuits. In 1978 the FTC opened public hearings about a possible rulemaking known as KidVid. The agency requested public comment on a plan to ban all advertising for children too young to understand the nature of commercials. For children 8–11, the plan was to ban advertising for all sugared products; for older kids, the plan was to introduce prohealth public service announcements to counter the claims of the sugared food advertisers. The FTC also proposed banning host selling and restricting ads that equated sugar with fun. Drawing heavily on a 1997 report from the National Science Foundation, the FTC produced its own report arguing that food advertising on children's television is both unfair and deceptive because children do not have the cognitive capabilities to understand the persuasive intent of advertising. The potential harm to children cited in the KidVid hearings was not obesity but instead the likelihood of dental caries from sweetened foods and beverages. At the time, half of American children by age 2 already had gum disease and at least one cavity. (For a retrospective review of KidVid, see Westen 2006; Pomeranz 2010; Wilde 2009.)

There was fierce opposition from the food, toy, broadcasting, and advertising industries to this potential rulemaking. Farmers were also opposed, as was the tobacco industry, which was concerned that regulation of food advertising would set a precedent for tobacco advertising. (Cigarette ads were already banned on television and radio, as of 1971. However, smokeless tobacco ads aired until 1986. During the 1970s, 1980s, and most of the 1990s, there was still tobacco advertising in magazines and newspapers, and on billboards.) The FTC received over 60,000 written comments and the transcripts of the hearings ran to over 6,000 pages. These industries argued that they had a First Amendment right to advertise as a form of commercial speech. The FTC had trouble providing strong scientific evidence that demonstrated a causal relationship between the advertising and the likelihood of cavities. The FTC also had trouble deciding on which specific products should be banned from advertising to children (e.g., potato chips and dried fruit are more cariogenic than candy), and the decision concerning which programs should be subject to advertising restrictions was difficult because children watched not only cartoon programs but also family programs that had a mixed audience of adults and children. For example, shows such as *I Love Lucy* and *The Andy Griffith Show* had large numbers of children viewers.

The industries were successful in their lobbying efforts. Registering its displeasure with the FTC's proposed regulation, Congress refused to approve the FTC's entire operating budget. However, the FTC stood firm in its plans, which led Congress eventually to pass in 1980 the sardonically named FTC Improvement Act. It essentially eliminated the Commission's rights to regulate unfair practices, although it left in place its right to bring *post hoc* suits against deceptive practices (Story and French 2004; Pomeranz 2010; Alderman et al. 2007; Mello 2010). The Act specifically forbade the FTC from taking any strong measures to restrict children's television advertising. Based on these actions and the appointment by President Reagan of a new FTC Commissioner who was not sympathetic to a strong regulatory hand by the Agency, the KidVid rulemaking initiative was abandoned in 1981. The practical outcome was to stop not only the FTC, but also every other federal agency from being a regulator of advertising on children's programming for the next 20 years.

The FTC Improvement Act was the product of the changing political sentiment in the United States against federal regulation. Antiregulatory sentiment was a hallmark of the Reagan administration. The attitude was that individual action was a personal responsibility (Alderman et al. 2007). If an individual ate unhealthy things or did not eat in moderation, it was the individual's fault and not the responsibility of the federal government. An influential editorial ran on March 1, 1978 in the *Washington Post* that captured this sentiment well, calling the FTC the "National Nanny." (For a discussion of the FTC in the 1970s and a reprint of the editorial, see Pertschuk 1982.)

3.2 Self-Regulation by the Food and Advertising Industries

If the federal government was not going to regulate food advertising to children, who would? The answer for more than 20 years was that industry regulated itself through an organization called the Children's Advertising Review Unit (CARU). In 1924 the American Association of Advertising Agencies approved a Creative Code that prohibited its members from knowingly writing advertising copy that contains: "false or misleading statements or exaggerations, visual or verbal; testimonials that do not reflect the real opinion of the individual(s) involved; price claims that are misleading; comparisons that unfairly disparage a competitive product or service; and statements, suggestions or pictures offensive to public decency or minority segments of the population" (AEF 2005). In 1971, in the face of public concern about the truthfulness of advertising and the possibility of federal regulation, the advertising industry and the CBBB banded together to create a system for national voluntary self-regulation of the advertising industry known as the National Advertising Review Council (renamed in 2012 as the Advertising Self-Regulatory Council) (American Association of Advertising Agencies 1990). The system originally included a National Advertising Division that provided a system for processing complaints about specific advertising campaigns without the time and expense of litigation, a National Advertising Review Board that provided an appeal process when the challenger or advertiser did not agree with an NAD decision, and Local Advertising Review Panels. CARU was added to this system in 1974, the Electronic Retailing Self-Regulation Program was added in 2004, and for online behavioral advertising the Online Interest-Based Advertising Accountability Program was added in 2011 (AEF 2005; ASRC 2012).

In 1973, several members of Congress asked FTC chairman Michael Pertschuk to investigate and possibly regulate television advertisements directed at children. To stave off federal regulation, the National Advertising Review Council created CARU, under the administration of the CBBB. CARU's policies covered children's magazines and children's television programs, but not family television despite the fact that many children watched these family programs. The rules, set by the CBBB and three advertising trade associations, provided guidelines that were intended to assure that these ads were not untruthful, misleading, or inappropriate. CARU based these guidelines, for example, about product claims or sales pressure, on the fact that children had limited cognitive skills in order to understand and process ads. (For a general discussion of CARU, see Story and French 2004; Alderman et al. 2007; Fried 2006; Armstrong 1984.)

An independent review of CARU was taken by a business school professor (Armstrong 1984). He found that the investigative casework at CARU was light. During its first 28 years, CARU reviewed a total of approximately 150 ads for food products and services (less than 15 % of all the ads it reviewed) and only ruled formally on 57 food ads—about 2 per year. The level of activity increased or decreased over time—corresponding to external pressures—reaching a high point of activity in 1978, when there was threat of FTC regulation. The staff (four people) and budget (just over $100,000 per year) were too meager to carry out the investigative function adequately. The funding came entirely from industry, and CARU was beholden only to industry. CARU has no strong sanctions to apply to companies that broke its advertising policies. Armstrong argued that CARU would have been more effective if, instead of focusing on investigative functions, it had spent more of its funding on the development of better advertising guidelines, seminars to teach advertisers how to advertise properly, clearinghouses of research on advertising and children, and review of ads before they appear on television—all of which CARU did to some limited extent in its early days.

Other critics noted additional problems. The CARU policies were lax, for example, allowing cartoon characters to appear in advertising on other shows, only being prohibited by CARU regulations from appearing in ads during the program in which the character appeared, and allowing claims to be made about the fact that a nutritionally depleted sweetened cereal was part of a nutritious breakfast so long as the image showed the cereal in a setting that included other foods that would, together with the cereal, constitute a nutritious breakfast. The CARU regulations, which focused on issues of accuracy and deception in specific ads, overlooked the fact that the massive number of ads in itself created an environment that had a strong influence on children (Byrd-Bredbenner 2002).

Critics such as Action for Children's Television and the Council on Children, Media, and Merchandising believed CARU did not go far enough, and they continued to push for federal regulation. In fact, the creation of CARU was not sufficient to prevent the FTC KidVid regulatory hearings, but once Congress stopped the FTC's regulatory efforts, CARU was the only regulatory body that remained. Many of the largest food companies, such as McDonald's, General Mills, and Hershey, became long-standing supporters of CARU.

Over time, the process of regulating food marketing to children became more complex than it had been at the time of KidVid and CARU's creation. Marketers began to employ child psychologists and apply more nuanced scientific findings to make their messages to children more effective. For example, Kids as Customers: A Handbook of Marketing to Children (McNeal 1992) shows how to apply findings from developmental psychology to more effectively persuade children. A secondary industry developed to provide this marketing expertise to the food companies. For example, the Geppetto Group, a New York-based marketing firm that is part of the British advertising conglomerate WPP Group, taught CocaCola, McDonald's, Frito-Lay, Kraft, and other clients how to improve their persuasiveness through the use of applied psychology and anthropology. In a counter to commercial use of psychology, Harris et al. (2009b) called for new psychological research as the basis for what they call a "food marketing defense model" aimed to counter harmful food marketing by better understanding the nature of awareness, understanding, ability, and motivation to resist this marketing.

In a development that sociologist Juliet Schor calls the commercialization of childhood, companies have instituted multipronged marketing campaigns to teach children to become life-long consumers, especially of particular national brands (Schor 2005; also see Hill 2011; U.S. General Accounting Office 2000). Advertising at first appeared in a single medium such as a television advertisement or on a

Television advertisement	Dexter's Lab GoGurt characterized as a scientific invention
Special packaging of food products	Pez containers with cartoon characters on top
Creation of licensed figures or games for branded foods	Sneak King advergame, featuring the Burger King king
Product placements in movies and comic strips	Campbell's Soup and Goldfish crackers in Alvin and the Chipmunks
Giveaways (also known as premiums)	toys in McDonald's Happy Meals
Character licensing from movies for new food products	Superman Crunch breakfast cereal
Branded books and toys	Oreo Cookie Counting Book
Other branded products	Barbie dressed in a McDonald's uniform
Promotional tours	Nabisco Nilla Wafers banana pie eating contests at various theme parks
Peer-to-peer marketing	Proctor and Gamble's Tremor marketing arm, which has recruited 240,000 young people to promote its products in everyday settings
Product fan clubs	Burger King Kids Club with its five million members
Email lists of people interested in a particular brand or product	Jack in the Box Connect
Exclusive selling rights in public schools	Pouring rights for PepsiCo products
Incentive programs linked to educational activities	McSpellit Club in which the student receives free or discounted food for getting good scores on spelling tests
Advertisements on Channel One, the national in-school current events program	For Sunny D sweetened orange drink
Food-company sponsored curricula such as using their products in nutrition and science materials handed out in schools	California high schools sing videos, DVDs, and worksheets created by Kraft and the egg, pork, and beef councils
Interactive food company websites	McDonald's McVideogame (http://www.mcvideogame.com/index-eng.html)

Fig. 3.3 Modern multiplatform food advertising to children. Source: Schor and Ford 2007; Grigorovici and Constantin 2004; Montgomery and Chester 2009; van Reijmersdal et al. 2010; Metrock 2013; Samuels and Associates 2006. Also see Linn and Golin 2006; and Linn and Novosat 2008

billboard, and regulatory efforts were made to keep advertising campaigns one-dimensional (e.g., no program characters appearing in the advertisements). But increasingly, industry used a coordinated effort to approach children simultaneously through multiple media, multiple techniques, and sophisticated psychological means. Figure 3.3 presents examples of the many elements of this modern marketing technique. (It is beyond the scope of this chapter to examine regulation and self-regulation of advertising to children using media other than television and the Internet. Some examples of literature that discuss this issue include Beales et al. 2004; Rotfeld and Parsons 1989; and Wilde 2009.)

Children sometimes understood what the food companies are trying to do with their advertising. For example, in a focus group conducted by academic researchers with children about what the food companies might do in order to sell more cereal, one child answered: "I would make my cereals addictive, I would cover all of the cornflakes in nicotine and then people would be addicted and then they would have to buy more" (Hill and Tilly 2002). It is of course not clear how indicative this one child is of the understanding of advertising by children in general; and understanding the concept of advertising does not mean being able to resist it, or even of recognizing it in its many different forms.

3.3 A New Public Concern About Childhood Obesity

While there was not much action on reining in television food advertising to children during the 1980s and 1990s, the issue received new public attention in the first decade of the new century. Part of the stimulus was the Surgeon General's pronouncement in 2001 about a childhood obesity epidemic in America, but in fact in this first decade of the twenty-first century a number of nonprofit and government groups began to pay attention to this problem and take action. In 2004, the American Psychological Association published a study about advertising and children, focused on the difficulties that children have in both recognizing and evaluating persuasive claims by advertisers (Kunkel et al. 2004). The report expressed concern about online advertising, unsupervised viewing of television advertising (with the increase of television sets in children's bedrooms), and advertising in schools (e.g., product placements in textbooks and commercials on Channel One school newscasts). It identified two cognitive skills essential for children to comprehend ads: the ability to differentiate ads from programming, which they report that 4- and 5-year-olds are typically not able to do; the ability to understand the persuasive intent of ads and evaluate claims made in ads, which the study argues are cognitive skills typically not yet present in most 7- or 8-year-olds. Some concepts appearing in ads the study found to be beyond the cognitive abilities of most children in the target audience—concepts such as "part of a balanced diet." What makes the situation more troublesome, the study argued, is that the ads are effective, especially when they involve premiums or celebrity or cartoon characters. More than half of children could recall a product after only a single viewing of an ad. The authors also mentioned other problems created by the advertising, such as increasing parent–child conflict and promotion of materialistic attitudes in children. Prosocial, public service announcements are often of limited effectiveness, the study argued, because these ads do not appear frequently enough on television.

It was not only the American Psychological Association that raised concerns. In 2004 the Henry J. Kaiser Family Foundation produced a report on the increasing use of media by children, suggesting media's role in the obesity crisis (Kaiser Family Foundation 2004). That same year, the CSPI, which had taken an interest in this advertising issue since the 1970s, called into question seven advertisements in the magazine *National Geographic Kids*. CARU admitted that five of these ads were in violation of its principles but had no sanctions to levy against the advertisers. CSPI called for food ads to be banned from children's television and magazines. The Campaign for a Commercial-Free Childhood, organized in 2004 out of the Harvard University Judge Baker Children's Center, targeted specific food manufacturers including General Mills, Nabisco, and Post to ban food advertisements to children. Two years later, the American Academy of Pediatrics urged its members to contact Congress about restricting these advertisements (American Academy of Pediatrics 2006; Grimes 2008).

But it was the 2005 report by the Institute of Medicine, *Food Marketing to Children and Youth: Threat or Opportunity?* that finally stimulated action (McGinnis et al. 2006; Wilde 2009). The Institute's report provided a comprehensive review of the scholarly literature (123 articles) on children, food, and advertising. Its analysis showed that the research literature provides strong evidence of the impacts of television advertising, although less strong evidence concerning the impact of other marketing media such as company Internet websites; also, strong evidence for "proximate outcomes" such as children's attitudes about foods and food choices, but less strong evidence about long-term weight gain. The report concluded that these marketing practices are not consonant with healthy diets for kids. Understanding the FTC's earlier political debacle with KidVid and the continuing political uneasiness with federal regulation, the Institute called on the food and advertising industries to self-regulate, but called for the government to step in if industry had not acted effectively after 2 years. The report also called upon the FTC to monitor the food industry's self-regulation efforts.

As public interest in childhood obesity as a public health issue increased, the prestigious Institute of Medicine began to pay more attention. While the Institute typically issued between 50 and 100 reports each year, it had seldom studied childhood obesity in the past. There had been reports on diet issues related to the federal Women, Infants, and Children program in 2000, 2002, 2004, and 2005, and a report on weight management in the military in 2003, but there were no general studies of childhood obesity in the decade prior to 2005. However, in the same year that the food marketing report came out, the Institute issued a report entitled *Preventing Childhood Obesity: Health in the Balance*, which made wide-ranging recommendations for action by local, state, and federal governments; industry and media; healthcare professionals; community and nonprofit organizations; schools; and parents. The Institute released another report the same year, *Preventing Childhood Obesity: Life in the Balance* (Koplan et al. 2005).

In response to this latter report, the Robert Wood Johnson Foundation asked (and funded) the Institute to assess various childhood obesity prevention programs, which resulted in three additional Institute of Medicine (2006a, b, 2007) reports: *Progress in Preventing Childhood Obesity: Focus on Schools*, *Progress in Preventing Childhood Obesity: Focus on Industry*, and *Progress in Preventing Childhood*

Obesity: How Do We Measure Up? Later in the decade, the Institute of Medicine (2008, 2009, 2010) published three additional reports related to childhood obesity: *Nutrition Standards and Meal Requirements for National School Lunch and Breakfast Programs, The Public Health Effects of Food Deserts*, and *School Meals: Building Blocks for Healthy Children.*

The Robert Wood Johnson Foundation established its own program in 2008, entitled Healthy Kids, Healthy Communities (Ohri-Vachaspati et al. 2012). The program's goal was to follow up on the initial *Preventing Childhood Obesity* report and address obesity by changing local policies and environments. The grants made to communities under this program so far have focused on giving incentives to retailers to sell healthier foods in food deserts, making purchase of foods from farms more viable, improving access to public outdoor recreational facilities, and enhancing the infrastructure for walking and cycling. It is too early to evaluate the overall success of this program.

Perhaps in response to the Institute's Food Marketing report, Kraft Foods surprised the rest of the food industry in 2005 by announcing that it would create no new television advertisements for certain lower nutrient foods in its product line and that it would phase out other marketing communications of these products (Page and Brewster 2007; Ellison 2005; Simon 2006 is highly critical of Kraft's initiative). This action angered some of Kraft's competitors, while public health advocates lauded the action (Wootan 2005). Kraft shifted its advertising budget targeted at children to more healthy alternative products such as sugar-free Kool-Aid. In the next several years, some of Kraft's competitors followed suit, mainly under the auspices of the Children's Food and Beverage Advertising Initiative (described below).

Industry took two collective actions in response to the new public interest in childhood obesity. It updated the CARU guidelines in 2006 and again in 2009. The new regulations included more rules about online advertising and provided clearer guidelines on issues concerning the blurring of programming and advertising. The revised guidelines also encouraged companies to display appropriate serving sizes in their ads and provide positive messages about balanced nutrition and healthy lifestyles (Grimes 2008; Mello 2010). The revisions were roundly criticized by CSPI for making few significant changes and being more interested in protecting marketers than in responding to the health needs of children and their parents (Jacobson 2006).

What CARU did not want to do was regulate which foods are deemed nutritious. In 2006, the CBBB and ten food companies formed the Children's Food and Beverage Advertising Initiative (Moore and Rideout 2007; Mello 2010). The goal was to shift the focus of advertising toward promoting healthier eating and healthier lifestyles. The Initiative provided self-regulatory, voluntary guidelines on which foods should be advertised to children under age 12. These included regulations to ensure that games on company websites would reinforce healthy eating and healthy lifestyles. The rules were written in a way that made it easy for food companies to comply with the guidelines: only 50 % of the ads from a company had to promote healthy eating and healthy lifestyles; at first, until the rule was changed under adverse public criticism, an ad counted as prohealth, even if it promoted an unhealthy product, if the ad also called for more exercise; and the companies decided for themselves which ads were targeted at children under the age of 12. The number of companies participating in the Initiative has grown from 10 to 17 and includes large industry players such as Kraft Foods, General Mills, CocaCola, PepsiCo, McDonald's, and Burger King. Some individual companies have taken individual action as well to provide more healthy products and to label them as such. For example, Disney has limited the licensing of its characters in the promotion of unhealthy foods (Mello 2010; Moore and Rideout 2007).

Researchers at Yale University's Rudd Center for Food Policy and Obesity (Schwartz et al. 2008, 2010) have analyzed the self-regulation under the Children's Food and Beverage Initiative. The main problem they identified was the wide variation in self-determination by individual companies of what constitutes a "better for you" product. The researchers noted that, under General Mills's criteria, Reese's Puffs, Cocoa Puffs, Lucky Charms, and Cookie Crisp—all General Mills cereals with high quantities of sugar—fall into the "better for you" category. The researchers measured the nutritive value of all 71 cereal brands they found advertised on U.S. television against the standards used by the British government (the United Kingdom Nutrient Profiling model) to determine whether a product has sufficient nutritive properties to be advertised on children's television in the United Kingdom. This model gives a single score to each product based on the totality of calories, fat, sugar, sodium, fiber, protein, and unprocessed fruits, nuts, and vegetables in the product. None of the General Mills cereals listed earlier and self-determined by the company to be "better for you" received high enough scores that they could have been advertised on British television. In fact, 98 % of the 71 cereal brands they tested received failing scores under the U.K. nutrient profiling. Kellogg's Pops Chocolate Peanut Butter and Quaker Cap'n Crunch's Crunch Berries received the lowest scores.

In order to better inform the Children's Food and Beverage Advertising Initiative, as well as the ongoing efforts of the federal agencies (FTC, HHS, and FCC), the Kaiser Family Foundation published a major study in 2007 under the title *Food for Thought*, which presented a comprehensive look at television food advertising to children (Gantz et al. 2007; U.S. Federal Trade Commission 2007). This study reviewed more than 1,600 h of television programming from 2005, including both shows directed at children and nonchildren's programming with a significant child audience. The study grouped children into three classes by age: 2–7, 8–12, and 13–17. It found that children of all ages were watching a great deal of television, but the 8- to 12-year-old group was watching the most. A typical 8–12 year old views more than 7,600 food ads—totaling more than 50 h of advertising—in a single year. Food ads are common on television, the study found, especially on children's programming where they constitute half of all ads. The ads are generally for low-nutrition foods—the most common being ads for candy and snacks (34 %), cereal (28 %), and fast food (10 %). There were no ads for fruits and vegetables. The ads most commonly appeal to taste (34 %), fun (18 %), and premiums and contests (16 %). They found overblown the public concern about the dangers of known figures from cartoons or movies selling less nutritious food, with spokes-characters appearing in only 10 % of the ads.

In response to calls from First Lady Michelle Obama, the White House Task Force on Childhood Obesity, the FTC, and others, in 2011 the Children's Food and Beverage Advertising Initiative adopted uniform nutrition standards to be used by all of its members, effective as of the end of 2013. The Initiative further agreed to upgrade its standards, as necessary, in light of new scientific information about nutrition such as the USDA's 2015 Dietary Guidelines for Americans. The uniform nutrition criteria are stronger than its previous standards. For example, under the new standards, there are specific limits within a given category on total calories, saturated fat, trans fat, sodium, and total sugars; and a food does not qualify solely on a claim of reduction ("25 % less sodium") or portion control ("100 calories in this package") (Better Business Bureau 2013).

At the same time that the nonprofits were writing reports and industry was rejiggering its practices of self-regulation, there was a call for renewed federal regulatory effort (Mello 2010). A 2004 Institute of Medicine report encouraged the FTC to use its authority and resources to ensure the compliance of industry self-regulation (Warner 2005). In response the following year, the FTC and the Department of Health and Human Services sponsored a public workshop on the marketing of food to children (Pomeranz 2010; Wilde 2009; Simon 2006). The political climate was not yet ready for increased government regulation to meet the growing public concern over childhood obesity. Both the House and Senate had Republican majorities, and in her opening remarks to the workshop Chairwoman of the FTC Deborah Platt Majoras made it clear that regulation of the marketing of specific types of foods was not something the FTC wanted. Majoras stated, "under the right circumstances, industry-generated action can address problems more quickly, creatively and flexibly than government regulation" (Warner 2005). Liberals, such as Senator Tom Harkin, criticized CARU for not solving the problem and for being "captive to the industry." By contrast, the Grocery Manufacturers Association used the workshop to present its plans to increase self-regulation guidelines by having its advertising review board appoint a task force to oversee advergames, expand the board's staff and budget, and add guidelines to monitor promotional tie-ins with video games and product placement in television shows (Warner 2005).

At the workshop, the agencies offered a set of general recommendations to industry: industry should (1) create new products that would lead to healthier eating; (2) use product packaging to control portion size; (3) improve on product labeling to identify calories and nutrients more clearly; (4) adopt minimal nutritional standards for the products they market to children; (5) disseminate more frequent and more effective prohealth messages; (6) tailor educational programs to reach racial and ethnic communities that have a higher incidence of childhood obesity; (7) broaden CARU's mission to include self-regulation of the Internet, interactive games, and other nontraditional forms of advertising; (8) expand the CARU advisory board to include more people with relevant expertise in such areas as nutrition and developmental psychology; (9) allow parents and others to file complaints with CARU; (10) ensure that the CARU staffing and resources are adequate to do the job well; and (11) find ways to deter violations, especially repeat violations, of the CARU guidelines. While all of these were good ideas, none of these recommendations led directly to change.

The workshop did not produce the empirical data that the sponsoring agencies had hoped for, and Congress called on the FTC to prepare a follow-up report on the growing frequency of childhood obesity and its connection to industry marketing to children (Federal Trade Commission and Department of Health and Human Services 2006). The FTC subpoenaed 44 food companies in the process of collecting information for its report. The FTC reported back on the enormous efforts to market unhealthy foods to children: in 2006, $870 million was spent on marketing to children and over a billion dollars on marketing to adolescents. Almost two-thirds of the advertising dollars were spent on carbonated beverages, fast-food restaurants, and breakfast cereals—with the rest spent mostly on other beverages, snack foods, and candy (U.S. Federal Trade Commission 2007). Although the most money was spent on television advertising, significant amounts were also spent on premiums, Internet websites, email and text messaging campaigns, packaging and in-store displays, celebrity endorsements, and product placements. (On packaging, see Berry and McMullen 2008; Gelperowic and Beharrell 1994; Hill and Tilly 2002; Kornblau 1961.) Cross-promotions, such as the use of licensed figures or tie-ins to movies, television programs, or toys, had become very common, the report noted. The report also found that most companies were following their self-regulation pledges, but the definition of what counted as healthy varied widely from company to company. The FTC recommended that the industry focus on four issues: healthy product development, appropriate portion sizes, in-school marketing, and presenting pronutrition messages (Botha et al. 2008).

Between 2006 and 2008, the FCC expressed interest in extending regulation of television advertising directed at children to both cable television and the Internet—the latter through restrictions on the display of food company Internet addresses (URLs) on television programming targeted at children. These intentions were connected to efforts in 2007 to provide the FCC with greater regulatory control over cable television, which had been weak under the provisions of the Cable Communications Act of 1984 (Dealbook 2007). No single action represented the plans better than the large fine charged to Univision for misclassifying certain children's programming as educational. However, intense industry lobbying caused the FCC to scale back its plans to regulate the cable television industry (Labaton 2007a, b; New York Times 2007; Oxenford 2007, 2008; Puzzanghera 2007).

In 2011 the FTC proposed new rules that would expand the COPPA to cover technological advances in smartphones, social networking, and geolocation (Silverman 2011). That same year the FTC also circulated for comment a draft report that called for prohibition of advertising of unhealthy foods on the broadcast media, the Internet, and other media through either regulation or self-regulation by 2016 (Oxenford 2011). At this time, neither the FCC not the FTC has gained strong regulatory control over children and the Internet in ways that would enable them to control food advertising.

One of President Obama's first legislative actions, his economic stimulus package known as the American Recovery and Reinvestment Act (passed in 2009), created the Interagency Working Group on Food Marketed to Children. This group, consisting of the FTC, Centers for Disease Control, Food and Drug Administration, and Department of Agriculture, was charged with studying and developing recommendations for food marketing aimed at people under 18 years old, identifying the nutritive content of food advertised, and determining the role the foods play in youth obesity. In 2011 the Working Group made its recommendations available for public comment. It recommended that by 2016 all food marketed primarily to people under age 18 should meet two basic nutrition principles: first, the foods should contribute to a healthy diet and include at least one of the following food types: fruits, vegetables, whole grain, fat-free or low-fat milk products, fish, extra-lean meat or poultry, eggs, nuts and seeds, or beans; second, the foods should not have a negative impact on health or weight (quantified by maximum amounts of saturated fats, trans fat, added sugars, and sodium they can contain).

Not surprisingly, the food, advertising, and media industries objected strenuously to these guidelines. The Sensible Food Policy Coalition, organized by the food industry, countered with a "Keep the Government out of your Kitchen" media campaign. The food industry called for leaving the regulation in the hands of the Children's Food and Beverage Advertising Initiative (Wilde 2009). Representative of the Conservative position, the Bernard Center for Women, Politics and Public Policy, directed by the conservative television commentator Michelle Bernard, questioned whether advertising increases consumption of food products to children, whether banning advertising can decrease obesity, and whether advertising is the most important causal factor in obesity. The Bernard Center made questionable claims that the Interagency Working Group's recommendations would increase food prices, have a disproportionate impact on low-income and minority families, and rob parents of useful information in making food choices for their families. The Bernard Center argued instead in favor of self-regulation (Bernard and Bradley 2011). Another and perhaps more powerful opponent of these guidelines was the Grocery Manufacturers Association, which has a proven track record of using lobbying and campaign contributions to fight both state and federal bills that restrict the sale of soda and junk food (Simon 2006). On the other side of this political debate were the CSPI and other public health groups that believe self-regulation is too lax and needs to be replaced with regulation. (On the problems with self-regulation, see Sharma et al. 2010; Simon 2006; Lang et al. 2006.) The Interagency Working Group recommendations did not make it through Congress intact. Who will win out in the long run is still to be determined, but industry has much deeper pockets than the public interest groups.

3.4 Food Advertising and First Amendment Rights

One of the familiar arguments lodged many times since the 1970s by the food, media, and advertising industries is that food advertising to children is protected commercial speech under the First Amendment. If the courts uphold wide First Amendment rights for commercial speech, it will be difficult for regulators to exert any meaningful control over food advertising to children. The U.S. Supreme Court first recognized the concept of commercial speech in 1942 in the case *Velentine v. Chrestensen*, even though the court ruled that there is no First Amendment protection for it (in this case to distribute handbills advertising paid admission to a First World War submarine docked in New York City).

The argument surfaced again during the summer of 2011 in response to the Interagency Working Group's recommendation. Viacom hired well-known Stanford University law professor Kathleen Sullivan to argue the industry's First Amendment position in testimony to the FTC. There is disagreement among legal scholars as to the constitutionality of these regulations on commercial speech (See, for example, Alderman et al. 2007; Graff 2008; Graff et al. 2012; Hertz 2002; Pomeranz 2010; Ramsey 2006; Redish 2011; Sullivan 2011; Wilde 2009). Those who argue that such regulation is legal do so on several grounds. If the argument holds that advertising to children is deceptive and misleading because children do not have the cognitive abilities to evaluate it, then deceptive and misleading commercial speech is not protected by the First Amendment. Moreover, if it is found that the advertising is about image (being cool, having fun) and has no informational content concerning the products, it is also not speech protected by the First Amendment. Scholars in favor of regulation also note there is precedent for having different First Amendment standards for adults and children (e.g., *Ginsberg v. New York*), and that First Amendment protections are often balanced against other public protections, in this case the health of the nation's children.

It is beyond the scope of this chapter to review all of the legal arguments, but the basics are important for understanding the legal debate. Through the early 1970s, the courts had not recognized First Amendment rights for commercial speech, but it was instead regarded as a regulable business practice. The case that gave commercial speech First Amendment protection was *Virginia State Board of Pharmacy v. Virginia Citizens Consumer Council* (1976). In this case, the court overturned a state ban on the advertisement of prescription drug prices because this advertising provides information that is important both to customers and to the efficient functioning of the marketplace.

The case that set the standards for regulation of commercial speech in light of the First Amendment rights determined in *Virginia Pharmacy* was *Central Hudson Gas and Electric Corp. v. Public Service Commission of New York* (1980), in which the court struck down a state regulation banning advertising by electric utilities. The court provided a four-part test to determine whether federal regulation of commercial speech is constitutional: (1) whether the speech concerns lawful activity and is not misleading; (2) whether the government interest is substantial; (3) whether the government regulation directly addresses the government interest; and (4) whether the regulation is the minimal one that could protect that interest.

The courts have been strict in their findings on the last two parts of the Central Hudson test. In a heavily cited case, *Lorillard Tobacco Company v. Reilly* (2001), the court overturned a Massachusetts law that banned advertising of cigarettes within 1,000 ft of schools because the law was not sufficiently narrow in its formulation. The best case for those who believe in regulation of food advertising appears to be to argue the case for the advertising being actually or inherently misleading, not just potentially misleading—based on the cognitive limitations of children to undertake three tasks: distinguish advertising from other content, recognize the persuasive intent of advertising, and have the grounds for evaluating an advertisement and acting on it. Industry has a strong argument against any sweeping regulation based on the fourth principle of the Central Hudson test. The situation with food companies advertising in schools is perhaps different because the courts treat schools as a protected space (Graff 2008; Pomeranz 2010; Mello 2010).

One academic study (Kline et al. 2006) suggests that those interested in regulating food advertising to children can learn lessons from the fight against childhood smoking and that it might be prudent to avoid the First Amendment issues altogether by employing alternate means of regulatory control. These include regulating a product directly (what, when, where, and how products are sold), regulating the retail sale of the product (e.g., age restrictions), employing land use regulations to limit where product retailers can operate, imposing product standards, or taxing or exacting a fee on the product. Additionally, one could regulate a product by agreement, through private or public binding contracts, private or public litigation settlements, or private or public nonbinding agreements. There is also an opportunity for government-sponsored education and counteradvertising.

3.5 Marketing Food to Children Through the Internet and Other Means

The Internet, which began to spread through the general public in the 1990s, has become a major force in shaping the strategies for marketing food to children (Story and French 2004). The Internet has two powerful advantages over television advertising: websites are interactive so that the children can playfully engage with a company's products, and the exposure times are much longer—multiple minutes on a website compared to a 30 s television ad—giving much greater opportunity for brand immersion (Moore and Rideout 2007). (On using the Internet for social good, namely, communicating public health messages, see Freeman and Chapman 2008.) Given that the Internet did not have the same long history of federal regulation as television and radio, the rise of food websites on the Internet opened a whole new front for battles between advertisers and regulators (Montgomery 2007).

One of the first studies of online advertising of food to children (Moore and Rideout 2007) found that most of the food companies that advertise to children on television also have websites. At the time of this study, these websites were receiving 49 million visits by kids annually, and that number can only have grown substantially over the past few years. Ninety percent of the sites then were for foods of low nutritional value. Brand benefit claims appeared four times as often as nutrition claims.

Food company websites often included games—often multiple games—for children to play. As the children became engaged in the play of the game, with its animation, music, and likeable characters, they were exposed peripherally but repeatedly to brands and logos—in a way that seemed to be effective at building brand awareness and loyalty. Many of the games were designed so that the children would want to play them again, e.g., to better their scores or compare their performance with the performance of other kids.

Many of these websites also employed peer-to-peer (also known as viral) marketing techniques, for example, the kids were encouraged to email their friends with a greeting card that included the company's logo or invited the friends to visit the website (Moore and Rideout 2007). The websites often included copies of the company's television ads, which the kids could view as many times as they wanted. The sites also commonly offered free items for the kids to take with them, such as screen savers with the company's logo or images of popular musical bands. Some offered sweepstakes as a way to entice the children to visit the web pages. Some offered special rewards, such as unlocking hidden games, to children who had made product purchases. Some sites offered product discounts. Many of the sites included tie-ins with movie or cartoon characters the children were familiar with.

There is no regulation of online advertising that compares directly to the restrictions on television advertising. For example, there are no limitations such as those in the Children's Television Act of 1990 on the amount of time that children may spend with ads, nor are there bumpers as there are on television that clearly separates programming from advertising. Moreover, the majority of the websites use electronic cookies to track the browsing behavior of individual children. The federal government has even had difficulty regulating more basic problems such as pornography online. The Children's Decency Act and the Child Online Protection Act, which protected children from exposure to indecent content online, were held to be unconstitutional on First Amendment grounds. The Children's Internet Protection Act protects children from harmful online content, but only during access in public schools and libraries. The one piece of federal legislation (described in detail later in this section) that has best protected children online has been the COPPA, which went into effect in 2000 and regulates the collection of information from children and the circumstances under which parental permissions are required by website operators. The FTC has enforced COPPA, fining companies that abused children's privacy on several occasions.

In particular, the FCC does not have authority over the Internet as it does over television. In 1997 CARU revised its self-regulation guidelines so as to include online advertising, but its Internet regulations were both weaker than the rules for television and were voluntary (Story and French 2004). One study (Weber et al. 2006) showed that only 9 of the 40 most popular children's food company websites followed the CARU guideline that required separation of programming from advertising.

Advergames are "computer games specifically created to function as advertisements to promote brands, where the entertainment content mimics traditional game forms" (Kretchmer 2005). In a study of the leading 100 national advertisers in the United States, more than half of the advergames were placed by food companies (Lee and Youn 2008). In one study of advergames (Culp et al. 2010), the researchers conducted a content analysis of websites advertised during a sample of 34 h of programming on two children's networks, the Cartoon Network and Nickelodeon, during 2006 and 2007. The researchers found 290 such websites. Eighty-four percent of the websites contained games, all included at least one brand identifier (a food product image, product package image, textual representation of a brand or company name, or most commonly a company logo), and there was on average one healthful message for every 45 exposures to brand identifiers. Other techniques for attracting kids to these sites included sweepstakes contests (47 % of the websites) and free downloads (36 %). Strategies to lengthen stay on the web page and hence increase exposure to the brand included many of the common tropes of videogames: play-again options (91 %), opportunity to work toward a higher level of game skill (52 %), the ability to post scores publicly so as to compete with others (32 %), and the ability to win free prizes (17 %). The authors found considerable blurring of the boundary between ads and programming, with few healthful messages to offset the commercial marketing. Because the games were interactive, children were typically exposed to the marketing material for 30 min, compared to 30 s on television. The authors estimated that approximately two billion ad viewings had been made on websites targeting children.

The popularity of advergames, as well as the scholarship on the topic, is growing rapidly. The general consensus is that advergames are proliferating and that they educate about brands not but not about health. The concern is that youth are targeted, that the games may be more persuasive than less popular traditional types of advertisements, and that young people do not have sufficiently well-developed judgment skills and are vulnerable in much the same way that they are vulnerable to television advertisements.

Branded entertainment is also becoming more popular as the effectiveness of traditional advertising formats declines (Lowrey et al. 2005). Besides brand placements in magazines (e.g., Van Reijmersdal et al. 2005), movies (e.g., Russell and Belch 2005), and television game shows (e.g., Gould and Gupta 2006), brand placements are beginning to show up in digital games (e.g., Garcia and Baker 2004).

The advergame format can be distinguished from "in-game" advertising. The latter more closely resembles traditional product placement, but within a game, whereas for an advergame, the game is specially made to promote the brand. An advergame is usually also less complex than a "real" game in which brands can be placed. Since advergames are rather simple in their design (no complex rules, short playing time, etc.), they can be easily distributed on different platforms, such as on websites, via email (tell a friend, viral marketing), on cell phones, and on interactive digital television (e.g., during a commercial break). Integrating brands into games is a growing business.

A number of public interest groups and scholars are fighting back against the power of advergames and other online advertising to influence children in the food area. The Media Education Foundation has produced videos for classroom use such as "Feeding Frenzy: The Food Industry, Marketing and the Creation of a Health Crisis" (Media Education Foundation 2013). The public interest group Campaign for a Commercial-Free Childhood holds protests, offers online resources, blogs, and sponsors education and advocacy conferences (Campaign for a Commercial-Free

Overview	Lee et al. 2009; Staiano and Calvert 2012
Other studies	Moore and Rideout 2007; Jain 2010; An and Stern 2011; Dahl, Eagle, and Baez 2009; Grimes 2008; Grossman 2005; Hernandez and Chapa 2010; Mallinckrodt and Mizerski 2007; Pempek and Calvert 2009; and Sukoco and Wu 2011; Henry and Story 2009; Bailey, Wise and Bolls 2009; Chang, Zhang, and Luo 2010; McIlrath 2006; Nairn and Dew 2007; Cauberghe and De Pelsmacker 2010; Lee and Youn 2008; Brustein 2009
On young children's recognition of brands and logos	Fischer et al. 1991

Fig. 3.4 Literature on advergames and online commercial practices concerning food and targeting children

Childhood n.d.). Media scholars Ian Bogost and Paulo Pedercini have satirized advergames using the medium of games (Terdimen 2006). To counter the criticisms of media directed at children, in 1983 the advertising industry created its own foundation, the Advertising Educational Foundation, which offers guidelines for self-regulation and produces and distributes educational content arguing for the value of advertising to society (Advertising Educational Foundation 2013). Figure 3.4 provides a list of some of the major scholarly literature on advergames and online commercial practices concerning food and targeting children.

Most of the analysis of Internet advertising of food to children has involved the study of the sites of specific food companies. However, Alvy and Calvert (2008) instead studied the ten websites most often visited by children 8–11 years old (not including ones that were primarily portals or mainly geared to adults) because children spend more time on these sites than they do on food company websites. The ten sites are Candyland.com, Neopets.com, Cartoonnetwork.com, Nick.com, Miniclip.com, Disney.com, Ebaumsworld.com, Barbie.com, Disneychannel.com, and Funnyjunk.com. Seven of the ten sites contained food advertising. The foods advertised on these sites were candy (248 instances), sweetened breakfast cereals (42), fast-food restaurants (9), chips (3), dairy products (3), and sweet snacks (1). The techniques used on these sites were similar to those used on television: attention-getting production features (animation, bold/colorful text, dynamic images), branded characters, and repetition. Several of these sites contained advergames and product placements.

In response to parental concerns about the private information that companies were collecting about their children and their children's online behavior, in 1998 Congress passed the Children's Online Privacy Protection Act, which restricted the data that companies could collect about children, effective in 2000. Enforcement of COPPA is the responsibility of the FTC. The law covers websites that are designed for children under age 13. It requires that a website have a privacy policy, give specific rules about when and how the company must obtain consent from the child's parent or guardian, and place restrictions on marketing to children under age 13.

There have been a few actions against well-known companies, such as Mrs. Fields Cookies and Hershey Foods, under this law. The largest fine was $1 million, imposed against the social networking website Xanga in 2006, for repeatedly allowing children under 13 to sign up without parental permission.

The FTC updated its COPPA regulations, effective July 2013. The update clarifies that commercial websites and online services directed at children under the age of 13 must get parental permission to collect geolocation information, identifies as a best practice that companies no longer collect photos or videos containing a child's image or an audio file with the child's voice unless they have gained parental permission, broadens the protection of screen names as personal information, and limits the use of persistent identifiers that enable one to track a user over time or across different websites or online services (BCP Business Center 2013).

There are a few studies of marketing to children in magazines. Magazines present a different set of challenges to those in television advertising, and in many ways are more similar to the issues that arise in Internet advertising. The boundaries between programming and advertising are more subtle in magazines than on television, especially because the advertisement can be hidden in editorials, comics, games, and puzzles. The amount of contact time with the magazine, as with the Internet, is also much longer than the ad contact on television. One small study of children in Australia indicated that children sometimes understand the persuasive content of magazine advertisement but at other times even older teenagers regarded items as informational rather than recognizing them to be the advertisements that they were. Many of the techniques were the same in magazines as online, such as promotions, character licensing, and games (Jones et al. 2010).

There have also been a few studies of food company advertising through the sponsorship of sports clubs and sporting events. In one study of New Zealand children ages 10–14, students had high recall of who the sponsors were for both their local sports club and their favorite professional sports franchise. The children had a favorable impression of the food company sponsor, and that favorable impression was increased either by receiving a voucher for food products for their good sports performance or by receiving a certificate for their sports activity (typically branded with the company name or logo). The children reported that they liked to "return the favor" to the company by buying their products for their having sponsored these sporting activities (Kelly et al. 2011a, b; also see Hoek and Gendall 2006).

3.6 Conclusions

While self-regulation has done some good to limit the types and amount of television food advertising to children, it has been inconsistently applied and generally does not go far enough. (For a national comparison of government regulation and industry self-regulation of television food advertising, consider the situation in Australia as reported in King et al. 2011 and Magnus et al. 2009.) The lobbying might of the food and advertising industries is great, and this has limited the ability of the federal government to regulate. However, there is also a public reluctance for federal regulation of what people do in their homes and how they apply their parenting skills. This was most acutely felt in the era of the Reagan presidency, when there was a backlash to the FTC characterized as the "national nanny." The uneasiness with federal regulation persists today.

Concerted national attention to an epidemic of childhood obesity, which can be defined as beginning with the Surgeon General's pronouncement in 2001, has continued through the Obama presidency. National attention has led to improved industry self-regulation, valuable work by health nonprofits such as the Henry J. Kaiser Family Foundation and the Robert Wood Johnson Foundation, persuasive reports from professional organizations such as the American Academy of Pediatrics and the Institute of Medicine of the National Academies of Science, and most recently by renewed federal government effort such as President Obama's Interagency Working Group on Food Marketed to Children.

As delivery and consumption of entertainment content shifts from televisions to personal computers and smartphones, so too does the nature of that content. Advergames are an increasingly popular form of advertisement and they are more effective than television commercials in the following ways: children will spend vastly more time with an advergame than they will watching a brief commercial; advergames are typically less expensive to produce and distribute than commercials; advergames may leverage social media technologies to encourage children to spread awareness of a product (viral marketing at no cost to the advertiser and then the marketing occurs inside the mediated space of another entity such as Facebook); advergames may be played away from the home, on mobile devices, in spaces where parental control or supervision is limited or impossible; the psychological factors that make game mechanics function easily allow for the introduction of persuasive elements; and advergames exist in the largely unregulated space of the Internet. So far, advergames are opposed primarily by private interest groups, not government regulators. For these various reasons, concern is reasonable that, over time, advergames will hold an increasingly important place in the enticement of children by commercial interests. Although the design and production on advergames is often of low quality, as compared to high-budget console games produced by large studios, it is conceivable that increases in quality and sophistication will occur as the sophisticated software packages used to create games become less expensive and as the interactive material itself plays an increasingly important role in marketing campaigns. It is similarly reasonable to believe that, to the extent that television commercials contribute to unhealthy practices in children, advergames have a negative effect on children's health. Advergames may actually have a more pronounced effect than television advertisements, due to the increased time a child spends engaged with a particular brand or product.

It is still too soon to know the final outcome of all these efforts, but the opportunities for a reduction in the number of obese children in the United States now seems possible. While progress is being made, the food industry must do more to employ meaningful nutritional standards in the foods they market to children. They must avoid regulatory loopholes and produce snacks that are truly healthy rather than ones that are less unhealthy than the products they replaced. They need to have sweeping advertising regulations that address the many different media (from television to the Internet to product placement) as well as the cross-media campaigns that are today being employed so effectively by the food industry. They must also obtain cooperation from media content providers to engage in marketing and advertising campaigns that support healthy foods for children. (These points were addressed by Jon Leibowitz, chairman of the FTC, at the 2009 workshop on Sizing Up Food Marketing and Childhood Obesity, see U.S. Federal Trade Commission 2009.)

References

Advertising Educational Foundation (2013) Advertising to children. http://www.aef.com/on_campus/classroom/speaker_pres/data/3005. Accessed 15 Oct 2013

AEF (2005) Industry regulations. http://www.aef.com/on_campus/classroom/speaker_pres/data/3006. Accessed 24 Oct 2013

Alderman J, Smith JA, Fried EJ, Daynard RA (2007) Application of law to the childhood obesity epidemic. J Law Med Ethics 35(1)90–112

Alvy LM, Calvert SL (2008) Food marketing on popular children's web sites: A content analysis. J Am Diet Assoc 108(4):710–713

American Academy of Pediatrics (2006) Children, adolescents, and advertising. Policy Statement, Committee on Communications. Pediatr 118(6):2563–2569

American Association of Advertising Agencies (1990) Standards of practice of the American Association of Advertising Agencies. https://ams.aaaa.org//eweb/upload/inside/standards.pdf. Accessed 24 Oct 2013

An S, Stern S (2011) Mitigating the effects of advergames on children: Do advertising breaks work? J Advert 40:43–56

Anderson P, Bruijn A, Angus K, Gordon R, Hastings G (2009) Impact of alcohol advertising and media exposure on adolescent alcohol use: A systematic review of longitudinal studies. Alcohol Alcohol 44(3):229–243

Armstrong GM (1984) An Evaluation of the Children's Advertising Review Unit. J Public Policy Mark 3:38–55

ASRC (2012) Supporting advertising industry self-regulation. http://www.asrcreviews.org/supporting-advertising-industry-self-regulation/. Accessed 24 Oct 2013

Bailey R, Wise K, Bolls P (2009) How avatar customization affects children's arousal and subjective presence during junk food-sponsored online video games. CyberPsychology and Behav 12:277–283

Barlett CP, Anderson C (2007) Violent Video Games and Public Policy. Center for the Study of Violence, Iowa State University, Ames, IA.

Barry AE, Goodson P (2009) Use (and misuse) of the responsible drinking message in public health and alcohol advertising: A review. Health Educ Behav 37:288–303

Batada A, Seitz MD, Wootan MG, Story M (2008) Nine out of 10 food advertisements shown during Saturday morning children's television programming are for foods high in fat, sodium, or added sugars, or low in nutrients. J Am Diet Assoc 108(4):673–678

Bayer R, Kelly M (2010) Tobacco control and free speech – an American dilemma. N Engl J Med 362(4):281–283

BCP Business Center (2013). Complying with COPPA: Frequently asked questions. Federal Trade Commission. http://www.business.ftc.gov/documents/Complying-with-COPPA-Frequently-Asked-Questions. 24 Oct 2013, Accessed 25 Oct 2013

Becker-Olsen KL, Norberg PA (2010) Caution, animated violence. J Advert 39(4):83–94

Beales JH, Jennings CJ, Engle MK (2004) Advertising to kids and the FTC: A regulatory retrospective that advises the present. Federal Trade Commission. http://www.ftc.gov/speeches/beales/040802adstokids.pdf. 2 March, Accessed 29 Oct 2013

Bernard MD, Bradley AR (2011) A question of smart government and the interagency working group on food marketed to children. Bernard Center for Women, Politics, and Public Policy, Potomac

Berry B, McMullen T (2008) Visual communication to children in the supermarket context: Health protective or exploitative? Agric Hum Values 25:333–348

Better Business Bureau (2013) CFBAI's specific uniform nutrition criteria. http://www.bbb.org/us/storage/16/documents/cfbai/CFBAI%20Uniform%20Nutrition%20Criteria%20Fact%20Sheet%20-FINAL.pdf. June 2013, Accessed 25 Oct 2013

Bijvank MN, Konijn EA, Bushman BJ, Roelofsma PHMP (2009) Age and violent-content labels make video games forbidden fruits for youth. Pediatr 123:870–876

Blum A (2010) When 'more doctors smoked Camels': Cigarette advertising in the *Journal*. Soc Med 5(2):114–122

Botha S, Fentonmiller K, Jennings C, Johnson M, Young K, Hippsley H, Engle MK, Ippolito PM (2008) Marketing food to children and adolescents: A review of industry expenditures, activities, and self-regulation. Federal Trade Commission, Washington DC

Botvin GJ, Goldberg CJ, Botvin EM, Dusenbury L (1993) Smoking behavior of adolescents exposed to cigarette advertising. Public Health Rep 108(2):217–224

Brownell KD, Horgen KB (2004) Food fight. Contemporary Books, Chicago

Brustein J (2011) McDonald's makes subtle play for children online. Bits (blog), New York Times. April 20. http://bits.blogs.nytimes.com/2011/04/20/mcdonalds-makes-subtle-play-for-children-online. 20 April 2011, Accessed 17 Dec 2013

Buizjen M, Valkenburg PM (2003a) The unintended effects of television advertising: A parent-child survey. Commun Res 30(5):483–503

Buizjen M, Valkenburg PM (2003b) The effects of television advertising on materialism, parent-child conflict, and unhappiness: A review of research. Appl Dev Psychol 24:437–456

Buijzen M, Van Reijmersdal EA, Owen LH (2010) Introducing the PCMC model: An investigative framework for young people's processing of commercialized media content. Commun Theory 20:427–450

Byrd-Bredbenner C (2002) Saturday morning children's television advertising: A longitudinal content analysis. Fam Consum Sci Res J 30(3):382–403

Byrd-Bredbenner C, Grasso D (2000) Trends in US prime-time television food advertising across three decades. Nutr Food Sci 30(2):59–66

Campbell A (1999) Self-regulation and the media. Fed Commun Law J 51(3):711–772

Campbell A (2006) Restricting the marketing of junk food to children by product placement and character selling. Loyola Los Angel Law Rev 39:447–506

Campaign for a Commercial-Free Childhood (nd) About CCFC. http://www.commercialfreechildhood.org/about-ccfc. Accessed 15 Oct 2013

Capella ML, Taylor CR, Webster C (2008) The effect of cigarette advertising on consumption: A meta-analysis. J Advert 17(2):7–18

Carter OBJ, Mills BW, Donovan RJ (2009) The effect of retail cigarette pack displays on unplanned purchases: Results from immediate postpurchase interviews. Tob Control 18:218–221

Cauberghe V, Pelsmacker PD (2010) The impact of brand prominence and game repetition on brand responses. J Advert 39(1):5–18

Celebucki CC, Diskin K (2002) A longitudinal study of externally visible cigarette advertising on retail storefronts in Massachusetts before and after the master settlement agreement. Tob Control 11(Supplement II):ii47–ii53

Center for Science in the Public Interest (2003) Pestering parents: How food companies market obesity to children. Center for Science in the Public Interest, Washington DC

Chang J (2010) Rated m for mature: Violent video game legislation and the obscenity standard. J Civ Rights Econ Dev 24(4):698–730

Chang Y, Yan J, Zhang J, Luo J (2010) Online in-game advertising effect: Examining the influence of a match between games and advertising. J Interact Advert 11:63–73

Charlesworth A, Glantz SA (2005) Smoking in the movies increases adolescent smoking: A review. Pediatr 116(6):1516–1527

Chernin A (2008) The effects of food marketing on children's preferences: Testing the moderating roles of age and gender. Ann Am Acad Polit Soc Sci 615:102–118

Chou SY, Rashad I, Grossman M (2008) Fast-food restaurant advertising on television and its influence on childhood obesity. J Law Econ 51(4):599–618

Ciolli A (2007) Joe Camel meets YouTube: Cigarette advertising regulations and user-generated marketing. Univ Toledo Law Rev 39(1):121–127

Collier JE, Liddell P Jr, Liddell GJ (2008) Exposure of violent video games to children and public policy implications. J Public Policy Mark 27(1):107–112

Connor SM (2006) Food-related advertising on preschool television: Building brand recognition in young viewers. Pediatr 118:1478–1485

Cornwell TB, McAlister AR (2011) Alternative thinking about starting points of obesity: Development of child taste preferences. Appetite 56:428–439

Courtney B (2006) Is obesity really the next tobacco? Lessons learned from tobacco for obesity litigation. Ann Health Law 15:61–64

Croley S (2003) White House review of agency rulemaking: An empirical investigation. Univ Chic Law Rev 70:821–885

Culp J, Bell RA, Cassidy D (2010) Characteristics of food industry web sites and 'advergames' targeting children. J Nutr Educ Behav 42(3):197–201

Dahl S, Eagle LC, Baez C (2009) Analyzing advergames: Active diversions or actually deception. Young Consum: Insight and Ideas for Responsib Mark 10(1):46–59

Dealbook (2007) F.C.C. reshapes rules limiting media industry. New York Times. http://dealbook.nytimes.com/2007/12/19/fcc-reshapes-rules-limiting-media-industry/?_r=0. 19 Dec 2007, Accessed 29 Oct 2013

Desrochers DM, Holt DJ (2007) Children's exposure to television advertising: Implications for childhood obesity. J Public Policy Mark 26(2):182–201

Dhar T, Baylis K (2011) Fast food consumption and the ban on advertising targeting children: The Quebec experience. J Mark Res 48(5):799–813

DiFranza JR, Richards JW Jr, Paulman PM, Wolf-Gillespie N, Fletcher C, Jaffe RD, Murray D (1991) RJR Nabisco's cartoon camel promotes Camel cigarettes to children. J Am Med Assoc 266(22):3149–3153

DiRicco DN, Shadel WG (2007) "Gender differences in adolescents' responses to themes of relaxation in cigarette advertising: Relationship to intentions to smoke," Addict Behav 32:205–213

Eckard WE Jr (1991) Competition and the TV cigarette advertising ban. Econ Inquiry 29 (1): 119–133

Ellison S (2005) Small bites: Why Kraft decided to ban some food ads to children. Wall Street Journal. http://online.wsj.com/news/articles/SB113072235663483869. 31 Oct 2005, Accessed 18 Dec 2013

Estlund C (2005) Rebuilding the law of the workplace in an era of self-regulation. Columbia Law Rev 105(2):319–404

Federal Trade Commission and Department of Health and Human Services (2006) Perspectives on marketing, self-regulation, and childhood obesity. Federal Trade Commission and Department of Health and Human Services, Washington DC

Fischer PM, Schwartz MP, Richards JW Jr, Goldstein AO, Rojas TH (1991) Brand logo recognition by children aged 3 to 6 years. JAMA 266(22):3145–3148

Folta SC, Goldberg JP, Economos C, Bell R, Meltzer R (2006) Food advertising targeted at school-age children: A content analysis. J Nutr Educ Behav 38:244–248

Forsyth SR, Malone RE (2010) 'I'll be your cigarette—light me up and get on with it': Examining smoking imagery on YouTube. Nicotine Tob Res 12(8):810–816

Freeman B, Chapman S(2008) Gone viral? Heard the buzz? A guide for public health practitioners and researchers on how web 2.0 can subvert advertising restrictions and spread health information. J Epidemiol Community Health 62:778–782
Freeman D, Brucks M, Wallendorf M, Bolland W (2009) Youths' understandings of cigarette advertisements. Addict Behav 34:36–42
Fried E (2006) Assessing effectiveness of self-regulation: A case study of the Children's Advertising Review Unit. Loyola Los Angel Law Rev 39:93–138
Frieden TR, Dietz W, Collins J (2010) Reducing childhood obesity through policy change: Acting now to prevent obesity. Health Aff 29(3):357–363
Galst JP, White MA (1976) The unhealthy persuader: The reinforcing value of television and children's purchase-influencing attempts at the supermarket. Child Dev 47(4):1089–1096
Gantz W, Schwartz N, Angelini JR, Rideout V (2007) Food for thought: Television food advertising to children in the United States. Henry J. Kaiser Family Foundation, Menlo Park
Garcia D, Baker VB (2004) Advergaming gets consumers to play while advertisers pay. Gartner G2 Report. https://www.gartner.com/doc/482698/advergaming-gets-consumers-play-advertisers. 22 Jan 2004, Accessed 18 Dec 2013
Gelperowic R, Beharrell B (1994) Healthy food products for children: Packaging and mother's purchase. Br Food J 96(11):4–9
Gentry E, Poirier K, Wilkinson T, Nhean S, Nyborn J, Siegel M (2011) Alcohol advertising at Boston subway stations: An assessment of exposure by race and socioeconomic status. Am J Public Health 101(10):1936–1941
Givel M (2007) A comparison of the impact of U.S. and Canadian cigarette pack warning label requirements on tobacco industry profitability and the public health. Health Policy 83:343–352
Glaeser EL, Schliefer A (2003) The rise of the regulatory state. J Econ Lit 41:401–425
Goldfarb A, Tucker C, (2010) Advertising bans and the substitutability of online and offline advertising. Journal of Marketing Research doi:10.2139/ssrn.1600221
Goldstein AO, Fischer PM, Richards JW Jr, Creten D (1987) Relationship between high school student smoking and recognition of cigarette advertisements. J Pediatr 110(3):488–491
Gostin LO (2009) FDA regulation of tobacco: Politics, law, and the public's health. J Am Med Assoc 302(13):1459–1460
Gould SG, Gupta PB (2006) Come on down: How consumers view game shows and the products placed in them. J Advert 35(1):65–81
Graff SK (2008) First amendment implications of restricting food and beverage marketing in schools, Ann Am Acad Polit Soc Sci 615:157–166
Graff S, Kunkel D, Mermin SE (2012) Government can regulate food advertising to children because cognitive research shows that it is inherently misleading. Health Aff 31(2):392–398
Grier SA, Kumanyika SK (2008) The context for choice: Health implications of targeted food and beverage marketing to African Americans. Am J Public Health 98(9):1616–1629
Grigorovici DM, Constantin CD (2004) Experiencing interactive advertising beyond rich media: Impacts of ad type on brand effectiveness in 3D gaming immersive virtual environments. J Interact Advert 5:22–36
Grimes SM (2008) Kids' ad play: Regulating children's advergames in the converging media context. Int J Commun Law Policy 8:162–178
Grossman S (2005) Grand theft Oreo: The constitutionality of advergame regulation. Yale Law J 115:227–236
Gunter B, Hansen A, Touri M (2009) Alcohol advertising and young people's drinking. Young Consum 10(1):4–16
Hamilton JL (1972) The demand for cigarettes: Advertising, the health scare, and the cigarette advertising ban. Rev Econ Stat 54(4):401–411
Hanewinkel R, Isensee B, Sargent JD, Morgenstern M (2010) Cigarette advertising and adolescent smoking. Am J Prev Med 38(4):359–366
Hanson J, Yosifon D (2003) The situation: An introduction to the situational character, critical realism, power economics, and deep capture. Univ Pa Law Rev 152:129–346

Harris JL, Bargh JA, Brownell KD (2009) Priming effects of television food advertising on eating behavior. Health Psychol 28(4):404–413

Harris JL, Brownell KD, Bargh JA (2009) The food marketing defense model: Integrating psychological research to protect youth and inform public policy. Soc Issues Policy Rev 3:211–271

Harris JL, Pomeranz JL, Lobstein T, Brownell KD (2009) A crisis in the marketplace: How food marketing contributes to childhood obesity and what can be done. Ann Rev Public Health 30:211–225

Hastings G, Stead M, McDermott L, Forsyth A, MacKintosh AM, Rayner M, Godfrey C, Caraher M, Angus K (2003) Review of research on the effects of food promotion to children. Report to the food standards agency, Glasgow. Centre for Social Marketing, University of Strathclyde, Glasgow

Hawkins K, Hane AC (2000) Adolescents' perceptions of print cigarette advertising: A case for counteradvertising. J Health Commun 5:83–96

Hebden RT (2011) Tertiary student drinking culture, Facebook, and alcohol advertising: Collapsing boundaries between social life and commercialised consumption. Master of Arts in Psychology, Thesis. Massey University

Henriksen L, Feighery EC, Schleicher NC, Cowling DW, Kline RS, Fortmann SP (2008) Is adolescent smoking related to the density and proximity of tobacco outlets and retail cigarette advertising near schools? Prev Med 47:210–214

Henriksen L, Schleicher NC, Feighery EC, Fortmann SP (2010) A longitudinal study of exposure to retail cigarette advertising and smoking initiation. Pediatr 126(2): 232–238

Henry E, Story M (2009) Food and beverage brands that market to children and adolescents on the internet: A content analysis of branded web sites. J Nutr Educ Behav 41(5):353–359

Henry J. Kaiser Family Foundation (2004) The role of media in childhood obesity. http://kaiserfamilyfoundation.files.wordpress.com/2013/01/the-role-of-media-in-childhood-obesity.pdf. Accessed 4 Nov 2013. Henry J. Kaiser Family Foundation, Menlo Park

Hernandez MD, Chapa S (2010) Adolescents, advergames, and snack foods: Effects of positive affect and experience on memory and choice. J Mark Commun 16:59–68

Hertz LM (2002) Advertising regulation on the internet. Comput Internet Lawyer 19:18–26

Hill H, Tilly J (2002) Packaging of children's breakfast cereal: Manufacturers versus children. Br Food J 104(9):766–777

Hill JA (2011) Endangered childhoods: How consumerism is impacting child and youth identity. Media Cult Soc 33:347–362

Hillman AJ, Keim GD, Schuler D (2004) Corporate political activity: A review and research agenda. J Manag 30:837–857

Hoek J, Gendall P (2006) Advertising and obesity: A behavioral perspective. J Health Commun 11:409–423

Hoek J, Gifford H, Thomson G, Edwards R (2010) How do tobacco retail displays affect cessation attempts? Findings from a qualitative study. Tob Control 19: 334–337

Hofferth SL, Sandberg JF (2001) How American children spend their time. J Marriage Fam 63:295–308

Holak SL, Reddy SK (1986) Effects of a television and radio advertising ban: A study of the cigarette industry. J Mark 50(4):219–227

Holt DJ, Ippolito PM, Desrochers DM, Keeley CR (2007) Children's exposure to TV advertising in 1977 and 2004: Information for the obesity debate. Bureau of Economics staff report. Federal Trade Commission, Washington DC

Hunter RJ Jr, Lozada HR, Mayo A (2011) Censorship in the video game industry: Government intervention or parental controls? Univ Denver Sports Entertain Law J. http://www.law.du.edu/documents/sports-and-entertainment-law-journal/issues/09/Article3.pdf. Accessed 4 Nov 2013

Huston AC, Wright JC, Rice ML, Kerkman D, St. Peters M (1990) Development of television viewing patterns in early childhood: A longitudinal investigation. Dev Psychol 26(3):409–420

IMDB.com (2013) Consuming kids: The commercialization of childhood (2008) – Quotes. http://www.imdb.com/title/tt1337599/quotes. Accessed 12 Dec 2013

Institute of Medicine (2006a) Progress in preventing childhood obesity: Focus on industry. http://www.nap.edu/openbook.php?record_id=11614&page=1. Accessed 4 Nov 2013
Institute of Medicine (2006b) Progress in preventing childhood obesity: Focus on schools. http://books.nap.edu/openbook.php?record_id=11461. Accessed 4 Nov 2013
Institute of Medicine (2007) Progress in preventing childhood obesity: How do we measure up? Committee on Progress in Preventing Childhood Obesity. National Academies Press, Washington DC
Institute of Medicine (2008) Nutrition standards and meal requirements for national school lunch and breakfast programs. http://books.nap.edu/openbook.php?record_id=12512. Accessed 4 Nov 2013
Institute of Medicine (2009) The public health effects of food deserts. http://books.nap.edu/openbook.php?record_id=12623. Accessed 4 Nov 2013
Institute of Medicine (2010) School meals: Building blocks for healthy children. http://books.nap.edu/openbook.php?record_id=12751. Accessed 4 Nov 2013
Ippolito PM, Pappalardo JK (2002) Advertising nutrition and health: Evidence from food advertising, 1977–1997. Bureau of Economics staff report. Federal Trade Commission, Washington DC
Jacobson MF (2006) Weak new voluntary guidelines on advertising to kids designed to protect marketers, not parents and families. Center for Science in the Public. Interest. http://cspinet.org/new/200611141.html. Accessed 2 Oct 2013
Jain A (2010) Temptations in cyberspace: New battlefields in childhood obesity. Health Aff 29(3): 425–429
John, DR (1999) Consumer socialization of children: A retrospective look at twenty-five years of research. J Consum Res 26(3):183–213
Jones SC, Mannino N, Green J (2010) 'Like me, want me, buy me, eat me': Relationship-building marketing communications in children's magazines. Public Health Nutr 13(12):2111–2118
Jordan AB (2008) Children's media policy. Future Child 18(1):235–253
Kahn FS (2002) Bombing markets, subverting the rule of law: Enron, financial fraud, and September 11, 2001 Tulane Law Rev 76:1579–1643
Kelly B, Baur LA, Bauman AE, King L (2011) Tobacco and alcohol sponsorship of sporting events provide insights about how food and beverage sponsorships may affect children's health. Health Promot J Aust 22(2):91–96
Kelly B, Baur LA, Bauman AE, King L, Chapman K, Smith BJ (2011) 'Food company sponsors are kind, generous and cool': (Mis)conceptions of junior sports players. Int J Behav Nutr Phys Act 8:95–101
Kent MP, Dubois L, Wanless A (2012) A nutritional comparison of foods and beverages marketed to children in two advertising policy environments. Obes 20(9):1829–1837
Kenyota G (2008) Thinking of the children: The failure of violent video game laws. Fordham Intellect Prop Media Entertain Law J 18(3):785–815
King C III, Siegel M (2001) The master settlement agreement with the tobacco industry and cigarette advertising in magazines. N Engl J Med 345(7):504–511
King C III, Siegel M, Celebucki C, Connolly GN (1998) Adolescent exposure to cigarette advertising in magazines: An evaluation of brand-specific advertising in relation to youth readership. J Am Med Assoc 279(7):516–520
King L, Hebden L, Grunseit A, Kelly B, Chapman K, Venugopal K. (2011) Industry self-regulation of television food advertising: Responsible or responsive? Int J Pediatr Obes 6(2–2): e390–e398
Kline R, Graff S, Zellers L, Ashe M (2006) Beyond advertising controls: Influencing junk-food marketing and consumption with policy innovations developed in tobacco control. Loyola Los Angel Law Rev 39:603–646
Koplan JP, Liverman CT, Kraak VI, (eds) (2005) Preventing childhood obesity: Health in the balance. National Academies Press, Washington DC
Kornblau C (1961) Packaging and supermarkets: The package in the marketplace. American Marketing Association, Chicago

Kretchmer SB (2005) Changing views of commercialization in digital games: In-game advertising and advergames as worlds in play. DIGTAR conference Changing Views: Worlds in Play. www.gamesconference.org/digra2005/overview.php. June 2005, Accessed 18 Aug 2006

Krugman DM, King KW (2000) Teenage exposure to cigarette advertising in popular consumer magazines. J Public Policy Mark 19(2):183–188

Kunkel D (1991) Crafting media policy: The genesis and implications of the Children's Television Act of 1990. Am Behav Sci 35:181–202

Kunkel D, Watkins B (1987) Evolution of children's television regulatory policy. J Broadcast Electron Media 31(4):367–389

Kunkel D, Wilcox B, Cantor J, Palmer E, Linn S, Dowrick P (2004) Report of the APA task force on advertising and children. American Psychological Association, Washington DC

Labaton S (2007a) F.C.C. planning rules to open cable market. New York Times. http://www.nytimes.com/2007/11/10/washington/10cable.html?pagewanted=all. 10 Nov 2007, Accessed 29 Oct 2013

Labaton S (2007b) Cable industry wins compromise on F.C.C. plans. New York Times. http://www.nytimes.com/2007/11/28/business/media/28cable.html?adxnnl=1&pagewanted=all&adxnnlx=1383076810-NVd9N+6bFKqhEUPUG36Xvg. 28 Nov 2007, Accessed 29 Oct 2013

Lang T, Rayner G, Kaelin E (2006) The food industry, diet, physical activity and health. Centre for Food Policy, City University, London

Lawson C (1991) Guarding the children's hour on TV. New York Times. http://www.nytimes.com/1991/01/24/garden/guarding-the-children-s-hour-on-tv.html?scp=5&sq=Action%20for%20Children's%20Television&st=cse. 24 Jan 1991, Accessed 4 Oct 2013

Lee M, Choi Y, Quilliam ET, Cole RT (2009) Playing with food: Content analysis of food advergames. J Consum Aff 43:124–159

Lee M, Youn S (2008) Leading national advertisers' use of advergames. J Curr Issues Res Advert 30(2):1–13

Leifer AD, Gordon NJ, Graves SB (1974) 'Children's television' More than mere entertainment. Harvard Educ Rev 44(2):213–245

Linn S, Golin J (2006) Beyond commercials: How food marketers target children. Loyola Los Angel Law Rev 39:13–32

Linn S, Novosat CL (2008) Calories for sale: Food marketing to children in the twenty-first century. Ann Am Acad Polit Soc Sci 615:133–154

Livingstone S, Helsper EJ (2006) Does advertising literacy mediate the effects of advertising on children? A critical examination of two linked research literatures in relation to obesity and food choice. J Commun 56:560–584

Lowe MR, Butryn ML (2007) Hedonic hunger: A new dimension of appetite. Physiol Behav 91:432–439

Lowrey TM, Shrum LJ, McCarty JA (2005) The future of television advertising In: Kimmel A (ed) Marketing communication: Emerging trends and developments. Oxford University Press, New York, pp 113–132

Luke DA, Ribisi KM, Smith C, Sorg AA (2011) Family Smoking Prevention and Tobacco Control Act. Am J Prev Med 40(3):295–302

Magnus A, Haby MM, Carter R, Swinburn B (2009) The cost-effectiveness of removing television advertising of high-fat and/or high-sugar food and beverages to Australian children. Int J Obes 33(10):1094–1102

Malik VS, Schultze MB, Hu FB (2006) Intake of sugar-sweetened beverages and weight gain: A systematic review. Am J Clin Nutr 84:274–288

Mallinckrodt V, Mizerski D (2007) The effects of playing an advergame on young children's perceptions, preferences, and requests. J Advert 36:87–100

McCool J, Webb L, Cameron LB, Hoek J (2012) Graphic warning labels on plain cigarette packs: Will they make a difference? Soc Sci Med 74:1269–1273

McGinnis JM, Gootman JA, Kraak VI, (eds) (2006) Food marketing to children and youth: Threat or opportunity? National Academies Press, Washington DC

McIlrath M, (2006) Children's cognitive processing of internet advertising. Dissertation, University of California, Santa Barbara

McNeal JU (1992) Kids as customers: A handbook of marketing to children. Lexington Books, New York

Media Education Foundation (2013) Feeding frenzy: The food industry, marketing & the creation of a health crisis. Video. http://www.mediaed.org/cgi-bin/commerce.cgi?preadd=action&key=308. Accessed 15 Oct 2013

Mello MM (2010) Federal Trade Commission regulation of food advertising to children: Possibilities for a reinvigorated role. J Health Politics Policy Law 35(2):227–276

Metrock J (2013) Channel One News brings back junk food advertising. http://www.obligation.org/2013-04-10-channel-one-news-brings-back-junk-food-advertising. 10 April 2013, Accessed 31 Oct 2013

Monaghan S, Derevensky J, Sklar A (2008) Impact of gambling advertisements and marketing on children and adolescents: policy recommendations to minimise harm. J Gambl Issues 22:252–274

Montgomery KC (2007) Generation digital. MIT Press, Cambridge

Montgomery KC, Chester J (2009) Interactive food and beverage marketing: Targeting adolescents in the digital age. J Adolesc Health 45:S18–S29

Moore ES, Rideout VJ (2007) The online marketing of food to children: Is it just fun and games? J Public Policy Mark 26(2):202–220

Morrison MA, Krugman DM, Park P (2008) Under the radar: Smokeless tobacco advertising in magazines with substantial youth readership. Am J Public Health 98(3):543–548

Nadeau MÈ (2011) Food advertising directed at children. Quebec Coalition on Weight-Related Problems, Report. http://www.childhoodobesityfoundation.ca/files/files/AMorel_food_advertising_directed_at_children.pdf. Accessed 19 Dec 2013

Nairn A, Dew A (2007) Pop-ups, pop-unders, banners, and buttons: The ethics of online advertising to primary school children. J Direct Data Digit Mark Pract 9:30–46

Nelson JP (2006) Cigarette advertising regulation: A meta-analysis. Int Rev Law Econ 26(2): 195–226

Nelson JP (2010) Alcohol advertising bans, consumption, and control policies in seventeen OECD countries, 1975–2000. Appl Econ 42(7):803–823

New York Times (1992) Watchdog group for children's TV to disband. http://www.nytimes.com/1992/01/09/us/watchdog-group-for-children-s-tv-to-disband.html?scp=17&sq=Action%20for%20Children's%20Television&st=cse. 9 Jan 1992, Accessed 4 Oct 2013

New York Times (2007) Regulating cable. Editorial. http://www.nytimes.com/2007/11/27/opinion/27tue3.html?_r=0. 27 Nov 2007, Accessed 29 Oct 2013

Ohri-Vachaspati P, Leviton L, Bors P, Brennan L, Brownson RC, Strunk S (2012) Strategies proposed by healthy kids, healthy communities partnerships to prevent childhood obesity. Prev Chron Dis 9:100292. doi:10.5888/pcd9.100292

Oxenford D (2007) FCC initiates inquiry into children's television rules. Broadcast Law Blog. http://www.broadcastlawblog.com/2007/04/articles/fcc-initiates-inquiry-into-childrens-television-rules/. 21 April 2007, Accessed 29 Oct 2013

Oxenford D (2008) The regulation of TV programming for children – embedded and interactive advertising, violence, and ratings. Broadcast Law Blog. http://www.broadcastlawblog.com/2008/06/articles/the-regulation-of-tv-programming-for-children-embedded-and-interactive-advertising-violence-and-ratings/. 22 June 2008, Accessed 29 Oct 2013

Oxenford D (2011) FTC requests comments on guidelines for advertising unhealthy foods to children. Broadcast Law Blog. http://www.broadcastlawblog.com/2011/05/articles/ftc-requests-comments-on-guidelines-for-advertising-unhealthy-foods-to-children/. 1 May 2011, Accessed 29 Oct 2013

Page RM, Brewster A (2007) Frequency of promotional strategies and attention elements in children's food commercials during children's programming blocks in US broadcast networks. Young Consum 8(3):184–196

Pecora N (2007) Action for Children's Television (ACT). Encyclopedia of children, adolescents, and the media, vol 1. Sage Reference, Thousand Oaks
Pecora N, John P. Murray, and Ellen Ann Wartella. 2006. Children and Television: Fifty Years of Research. Routledge, London.
Pempek, T. A., and S. L. Calvert. 2009. "Tipping the Balance: Use of Advergames to Promote Consumption of Nutritious Foods and Beverages by Low-Income African American Children." Archives of Pediatr and Adolescent Medicine 163:633–637.
Pereira MA, Kartashov AI, Ebbeling CB, Van Horn L, Slattery ML, Jacobs DR Jr, Ludwig DS (2005) Fast-food habits, weight gain, and insulin resistance (the CARDIA study): 15-year prospective analysis. Lancet 365:36–42
Pertschuk M (1982) Stoning the national nanny: Congress and the FTC in the late 1970s. In: Revolt against regulation: The rise and pause of the consumer movement. University of California Press, Berkeley
Pettersson A, Olsson U, Fjellstrom C (2004) Family life in grocery stores: A study of interaction between adults and children. Int J Consum Stud 28(4):317–328
Pierce JP, Messer K, James LE, White MM, Kealey S, Vallone DM, Healton CG (2010) "Camel no. 9 cigarette-marketing campaign targeted young teenage girls". Pediatr 125(4):619–626
Pollay RW (1995) Targeting tactics in selling smoke: Youthful aspects of 20th century cigarette advertising. J Mark Theory Pract 3(1):1–22
Pollay RW, Siddarth S, Sigel M, Haddix A, Merritt RK, Giovino GA, Eriksen MP (1996) The last straw? Cigarette advertising and realized market shares among youths and adults, 1979–1993. J Mark 60:1–16
Pomeranz JL (2010) Television food marketing to children revisited: The Federal Trade Commission has the constitutional and statutory authority to regulate. J Law Med Ethics 38(1): 98–116
Powell Lisa M, Szczpka G, Chaloupka FJ (2007) Exposure to food advertising on television among US children. Arch Pediatr Adolesc Med 161:553–560
Powell LM, Szczpka G, Chaloupka FJ, Braunschweig CL (2007) Nutritional content of television food advertisements seen by children and adolescents. Pediatr 120:576–583
Preston E, White CL (2004) Commodifyingkids: Branded identities and the selling of adspace on kids' networks. Commun Q 52(2):115–128
Puzzanghera J (2007) FCC to fine Univision $24 million. Los Angeles Times. http://articles.latimes.com/2007/feb/25/business/fi-univision25. 25 Feb 2007, Accessed 29 Oct 2013
Quotes on the Media and Children (nd) http://www.zona-pellucida.com/media.html. Accessed 12 Dec 2013
Rabin RL (1986) Federal regulation in historical perspective. Stanford Law Rev 38:1189–1326
Ramsey WA (2006) Rethinking regulation of advertising aimed at children. Fed Commun Law J 58(2):367–398
Redish MH (2011) Childhood obesity, advertising and the first amendment. White Paper. Northwestern University School of Law. http://www.gmaonline.org/file-manager/Health_Nutrition/childhood_advertising__firstamendment.pdf. 8 June 2011, Accessed 27 Sept 2013
Richards TJ, Padilla L (2009) Promotion and fast food demand. Am J Agric Econ 91(1):168–183
Robinson TN, Borzekowski DLG, Matheson DM, Kraemer HC (2007) Effects of fast food branding on young children's taste preferences. Arch Pediatr Adolesc Med 161:792–797
Rose-Ackerman S (1990) Defending the state: A skeptical look at regulatory reform in the eighties. Univ Colo Law Rev 61:517–535
Rose-Steinberg J (2010) Gaming the system: An examination of the constitutionality of violent video game legislation. Seton Hall Legis J 35(1):199–220
Rosenheck R (2008) Fast food consumption and increased caloric intake: A systematic review of a trajectory towards weight gain and obesity risk. Obes Rev9:535–547
Rotfeld HJ, Parsons PR (1989) Self-regulation and magazine advertising. J Advert 18(4):33–40
Rubin L (2005) The myth of accountability and the anti-administrative impulse. Mich Law Rev 103:2073–2136

Russell CA, Belch M (2005) A managerial investigation into the product placement industry. J Advert Res 45(1):73–79
Samuels & Associates (2006) Food and beverage marketing of California high school campuses survey: Findings and recommendations. Commissioned by the Public Health Institute. http://ftc.gov/os/comments/foodmktgtokids-pra/526194-00019.pdf. March 2006, Accessed 31 Oct 2013
Sargen JD, Gibson J, Heatherton TF (2009) Comparing the effects of entertainment media and tobacco marketing on youth smoking. Tob Control 18(1):47–53
Scholzman KL (2011) Political participation. In: Kurian GT (ed) The encyclopedia of political science, vol 4. CQ Press: Washington DC
Schor J (2005) When childhood gets commercialized, can children be protected? Paper prepared for the Yale Legal Theory Workshop. http://www.yale.edu/law/leo/052005/papers/schor.pdf. Accessed 19 Dec 2013
Schor JB, Ford M (2007) From tastes great to cool: Children's food marketing and the rise of the symbolic. J Law Med Ethics 35(1):10–21
Schuler DA, Rehbein K, Cramer RD (2002) Pursuing strategic advantage through political means: A multivariate approach. Acad Manag J 45(4):659–672
Schwartz MB, Ross C, Harris JL, Jernigan DH, Siegel M, Ostroff J, Brownell KD (2010) Breakfast cereal industry pledges to self-regulate advertising to youth: Will they improve the marketing landscape? J Public Health Policy 31(1):59–73
Schwartz MB, Vartanian LR, Wharton CM, Brownell KD (2008) Examining the nutritional quality of breakfast cereals marketed to children. J Am Diet Assoc 108:702–705
Sebrie EM, Glantz SA (2007) Tobacco industry 'youth smoking prevention' programs to undermine meaningful tobacco control in Latin America. Am J Public Health 97(8):1357–1367
Shadel WG, Tharp-Taylor S, Fryer CS (2009) How does exposure to cigarette advertising contribute to smoking in adolescents? The role of the developing self-concept and identification with advertising models. Addict Behav 34(11):932–937
Sharma LL, Teret SP, Brownell KD (2010) The food industry and self-regulation: Standards to promote success and to avoid public health failures. Am J Public Health 100(2):240–246
Silverman D (2011) FTC proposes new rules for websites directed to children. Broadcast Law Blog. http://www.broadcastlawblog.com/2011/09/articles/ftc-proposes-new-rules-for-websites-directed-to-children/. 19 Sept 2011, Accessed 29 Oct 2013
Simon M (2006) Can food companies be trusted to self-regulate: An analysis of corporate lobbying and deception to undermine children's health. Loyola Los Angel Law Rev39:169–236
Spencer EH, Frank E, McIntosh NF (2005) Potential effects of the next 100 billion hamburgers sold by McDonald's. Am J Prev Med 28(4):379–381
Staiano AE, Calvert SL (2012) Digital gaming and pediatric obesity: At the intersection of science and social policy. Soc Issues Public Rev 6(1):54–81
Story M, French S (2004) Food advertising and marketing directed at children and adolescents in the US. Int J Behav Nutr Phys Act 1:3. doi:10.1186/1479-5868-1-3
Strasburger VC (2009) Media and children: What needs to happen? J Am Med Assoc 301(21):2265–2266
Sukoco BM, Wu W (2011) The effects of advergames on consumer telepresence and attitudes: A comparison of products with search and experience attributes. Expert Syst Appl 38:7396–7406
Sullivan KM (2011) The Interagency Working Group's Proposed Nutrition Principles to Guide Industry Self-Regulation Efforts: Constitutional Issues. Appendix A to Comments of Viacom Inc. before the Federal Trade Commission: In the Matter of Interagency Working Group on Food Marketed to Children: General Comments and Proposed Market Definitions. http://www.ftc.gov/sites/default/files/documents/public_comments/preliminary-proposed-nutrition-principles-guide-industry-self-regulatory-efforts-project-no.p094513-7884/07884-80045.pdf. Accessed 19 Dec 2013. Federal Trade Commission, Washington, DC.
Sung Y, Hennink-Kaminski H (2008) The master settlement agreement and visual imagery of cigarette advertising in two popular youth magazines. Journal Mass Commun Q 85(2):331–352

Terdimen D (2006) Games that stick it to 'the man'. La Molleindustria. http://www.molleindustria.org/node/149/. 1 Feb 2006, Accessed 15 Oct 2013

U.S. General Accounting Office (2000) Commercial activities in schools. Report No. GAO/HEHS-00-156

U.S. Federal Trade Commission (2007) Media and childhood obesity. Task force on media and childhood obesity. Federal Trade Commission, Washington DC

U.S. Federal Trade Commission (2009) Sizing up food marketing and childhood obesity. http://www.ftc.gov/bcp/workshops/sizingup/transcript.pdf. Accessed 4 Nov 2013. Federal Trade Commission, Washington DC

Uscinski HJ (1984) Deregulating commercial television: Will the marketplace watch out for children? Am Univ Law Rev 34:141–147

Van Reijmersdal EA, Jansz J, Peters P, Van Noort G (2010) The effects of interactive brand placements in online games on children's cognitive, affective, and conative brand responses. Comput Hum Behav 26:1787–1794

Van Reijmersdal EA, Neijens PC, Smith GE (2005) Readers' reactions to mixtures of advertising and editorial content in magazines. J Curr Issues Res Advert 27(2):39–53

Warner M (2005) Food industry defends marketing to children. New York Times. http://www.nytimes.com/2005/07/15/business/15adco.html?_r=0. 15 July 2005, Accessed 19 Dec 2013

Weber K, Story M, Harnack L (2006) Internet food marketing strategies aimed at children and adolescents: A content analysis of food and beverage brand web sites. J Am Diet Assoc 106(9):1463–1466

Westen T (2006) Government regulation of food marketing to children: The Federal Trade Commission and the Kid-Vid controversy. Loyola Los Angel Law Rev 39:79–91

Wilde P (2009) Self-regulation and the response to concerns about food and beverage marketing to children in the United States. Nutr Rev 67(3):155–166

Wojciechowski B (2010) The cultural, psychological, and legal aspects of video game censorship. Thesis, Honors College, University of South Florida

Wootan MG (2005) Kraft advertising-to-kids policy applauded. Center for Science in the Public Interest. http://www.cspinet.org/new/200501125.html. Accessed 2 Oct 2013

Zimmerman FJ, Bell JF (2010) Associations of television content type and obesity in children. Am J Public Health 100(2):334–340

Zimmerman FJ, Christakis DA (2005) Children's television viewing and cognitive outcomes: A longitudinal analysis of national data. Arch Pediatr Adolesc Med 159:619–625

Zuckerman DM, Zuckerman BS (1985) Television's impact on children. Pediatr 75(2):233–240

Chapter 4
American School Lunch Policy: A History

It is more fun to talk with someone who doesn't use long, difficult words but rather short, easy words like 'What about lunch?'

(Winnie-the-Pooh, Milne 1926)

Ensuring that every child is fed breakfast and lunch, as well as improving the nutritional value of the foods children eat at school, has been an important goal in the policy effort to control childhood obesity in the United States. This policy area received considerable public attention in 2001, when Surgeon General of the United States David Satcher declared childhood obesity to be a national epidemic (Jackson 2001; U.S. Department of Health and Human Services 2001; also see the Congressional testimony in 2003 by Richard Carmona, Satcher's successor as Surgeon General). Although one primarily identifies schools with education, schools are also important institutions for the health of American children. Not only do most children eat one or two meals there every school day, the schools also teach kids about nutrition and aid in developing healthy eating habits.

Federal programs have existed since the 1930s to feed America's children, and for many years these programs were seen as critical to eliminating hunger and malnutrition. However, in the past 20 years, food in the schools has been criticized for contributing to the obesity epidemic, both through what is served as part of the federally subsidized school lunch program and because of the competitive foods available to students from vending machines, snack bars, school stores, and as a la carte items in the school cafeteria. This chapter recounts the political history of the school lunch program and several closely related federal food programs in the United States. It considers informal policy activities by public interest groups, professional health organizations, and concerned parents and citizens as well as formal actions by federal, state, and local governments.

There are many actors in this story. Formal approaches have taken the form of laws passed by Congress (e.g., the Child Nutrition Act, the National School Lunch

W. Aspray et al., *Formal and Informal Approaches to Food Policy*, SpringerBriefs in Food, Health, and Nutrition, DOI 10.1007/978-3-319-04966-3_4,

Act) as well as reports and regulations produced by federal agencies such as the Department of Agriculture, the General Accounting Office, and the Department of Health and Human Services. There have also been formal policy efforts at the state and local levels, such as state taxation on sweetened beverages and policies set by local school boards regarding the presence, location, and content of vending machines. The courts have also been involved in setting formal policy, e.g., *State v. Whittle Communications* (North Carolina) on the rights of school boards to set commercial contracts. Informal policy efforts have included professional organizations (e.g., the Institute of Medicine, the American Academy of Pediatrics), public interest organizations (Consumers Union, Center for Science in the Public Interest), and concerned parents and other individuals (Jamie Oliver, Avis Richards, Sarah Wu). Opposing positions have been taken by food trade associations (e.g., the National Soft Drink Association, National Restaurant Association) as well as by individual companies in the fast food, food service, and beverage distribution industries.

This is a story about political and economic tensions and their resolution. These tensions include federal versus local control of the public schools, the economic battle between higher priced fresh foods and lower priced industrially produced foods, extra revenue for financially strapped schools from snack and soda sales versus a school environment promoting good health, and government regulation versus self-regulation by the food and beverage industry.

The literature on school lunch policy is growing. Numerous book-length accounts of the causes—sometimes presenting proposed solutions—to childhood obesity and the role of school lunches in this epidemic have been written in recent years by both advocates (Center for Science in the Public Interest 2003; Brownell and Horgen 2004; Critser 2004; Cooper and Holmes 2007; Kalfa 2011; Adamick 2012) and scholars (Nestle 2002; Levine 2010; Popendieck 2010).

4.1 Government Food Programs for Schools

In the nineteenth century, families were generally responsible for their children's school lunches. In the second half of the century, in a few large cities, private charitable organizations began to help with school lunches for the most needy children. For example, in 1853 The Children's Aid Society of New York started a program to feed students attending vocational school. The importance of good nutrition to the mental and physical well-being of children was spelled out in Robert Hunter's 1904 book, *Poverty*, which identified his experiences in the Chicago and New York slums, and in John Spargo's 1906 book, *The Bitter Cry of the Children*. As a result, the number of cities offering nutritious school lunches increased, and for the first time local governments or school boards began to assume responsibility for these lunches. For example, in 1908 Philadelphia's school board took over responsibility for the lunch program from charitable organizations.

By the 1920s, a number of schools were serving hot lunches, paid for by parent–teacher organizations, local charities, or local governments. During the Depression years of the 1930s, many more students came to school without the ability to provide or afford their own lunches, and the costs of providing them grew to a point where charities could not bear the expense. It became more common for local and state governments to pitch in. By 1937, 15 states had passed laws authorizing school boards to operate lunchrooms. In most cases the schools provided lunch at the cost of the food, but 4 states reduced these costs for children of poor families (Cummings 1940; Levine 2010). The first federal support came from the Reconstruction Finance Corporation in 1932, during the final year of the Hoover Administration. In 1935, FDR's Work Projects Administration assumed responsibility for school lunch support, and this support continued through the first several years of the Second World War. With the high level of wartime employment, by then most families could afford to feed their own children, while the food surplus on the farms that had been redistributed to the schools dwindled as the available crops were used to feed the military forces.

Once the war ended, however, there was once again concern about children receiving adequate nutrition at school. Senator Richard Russell from Georgia proposed a permanent lunch program, and in 1946 President Truman signed legislation creating the National School Lunch Program. It continues today, offering free or low-cost lunches to children from families of limited financial means who attend public schools, nonprofit private schools, or residential child-care facilities. The federal government subsidizes the schools to provide these lunches at reduced cost. The program was designed not only to help eliminate hunger and malnutrition among America's youth, but also to prop up farm prices in times of surplus farm production. In fact, Senator Russell was willing to sponsor this legislation primarily because of the economic boost it gave to southern farmers. The lunch program was operated as a food program rather than an education program—administered by the USDA rather than by the Department of Education. Both the meat and dairy industries profited substantially from this legislation.

During the 1950s, the lunch program mainly provided subsidized lunches to middle-class kids who could afford them. Senator Russell and the Southern Democrats insisted on local control of the programs even though they were carried out with federal funding. The lunch program was not mandatory, and as of 1960 only about half of the nation's schools participated. The number of free lunches was minimal, and many poor children (especially poor Black children) did not benefit from the program. In schools with free lunches, there was often stigmatization of the free-lunch children: separate food lines, different meals, and sometimes a requirement for the students to work for their lunches (Levine 2010).

During the Kennedy and Johnson administrations in the 1960s, education and welfare funds from the federal government were used in part to fund school lunch programs for the poor. It was a major element of President Johnson's War on Poverty. Nevertheless, there remained inequities in lunch programs between rich and poor school districts. The USDA still supported the school lunch program primarily because of its support for farmers, and the agency resisted efforts to tie the school lunch program more closely to the welfare program. In fact, the USDA did little to help the poor. Even the 1966 Child Nutrition Act, which promised every poor child a free lunch, left most of the control in local hands. However, the large Democratic victory in the national elections of 1964 brought many new liberals into Congress, and with them came increased support for the elimination of both poverty and racial injustice. By the end of the Johnson presidency, the school lunch program had been transformed from a farm subsidy program into a poverty program. This approach continued during the Nixon Administration, which provided additional funds to provide free lunch to children from poor families (Levine 2010).

Serving not only subsidized lunches to the middle class but also free lunches to America's poor children proved costly. As school systems scrimped to serve all of these children, the quality of meals declined and the number of middle-class children who were paying customers dropped precipitously. This created a new financial crisis for the program. In an interesting policy battle, antipoverty lobbyists and liberal politicians such as Senator George McGovern supported the privatization of the school lunch program as a means to lower unit costs and increase scale. Their main concern was that schools be able to provide free lunches to poor children. Thus, they sanctioned the introduction of fast food provided by such companies as McDonald's, Pizza Hut, and Taco Bell, as well as the use of processed foods from industrialized food service companies such as Sysco and Armour. This decision resulted in the increasing appearance in school cafeterias of prefabricated products, with a heavy reliance on frozen, dehydrated, and reheatable foods. The nutritionists lobbied unsuccessfully against privatization because they were concerned that the meals being supplied were not providing adequate nutrients to the children. The industrial food service providers managed to meet the minimal nutritional standards set by the USDA by fortifying with vitamins and minerals the foods they served. This set the stage for the nutritional problems of school lunches facing the nation at the turn of the new Millennium (Levine 2010).

In the 1980s, as part of its overarching efforts to downsize government and eliminate waste, the Reagan Administration proposed budget cuts that would have reduced federal funding for school lunches by almost 25 %. In order to qualify for free school lunches, as an antifraud measure, a family was required to supply the names and Social Security numbers of all adults who resided in the household. In a famous episode in 1981 that embarrassed the Reagan Administration, ketchup was designated as a vegetable in order to make it easier to meet the minimal nutritional standards. The American public was highly supportive of the school lunch program, and Congress did not go along with proposed cuts (Levine 2010). This is an example in which informal means, namely widespread public support, led to a formal continuation of the school lunch program.

The school lunch program has generally been considered a success, and it has survived through various economic downturns and political changes. In recent years, some public health advocates have criticized the program because the foods provided under the program are typically high in fat, salt, and calories. Despite all of these challenges, by the mid-2000s the program was serving over 30 million school children each school day.

While the school lunch program has been prominently in the public eye in recent years, the US Department of Agriculture operates five other programs that also contribute to food policy for the public schools (Alderman et al. 2007). The Child Nutrition Act, signed into law by President Johnson in 1966, established the School Breakfast Program as a supplement to the school lunch program. The breakfast program is open to all students at the same facilities as the school lunch program.

Breakfasts are available to all children at these schools, and the cost is reduced or eliminated for those children whose family incomes are sufficiently low. By the mid-2000s, this program was serving approximately eight million children every school day (USDA 2013b).

The Child and Adult Care Food Program, created in 1975 and extended to adults in 1987, serves low-cost or free meals to more than three million children and 100,000 elderly and disabled people each day. The federal government provides grants to individual states, and they in turn reimburse the costs for these programs at child-care centers, after-school centers, family daycare centers, and homeless and emergency shelters, as well as centers caring for elderly and disabled adults (USDA 2013a).

The Summer Food Service Program, created in 1975, provides meals to children in low-income areas during the summer months when school is out of session. In summer 2001, the program served 130 million meals to over 2 million students. The sponsors include public or private nonprofit schools, local and state governments, tribal governments, private nonprofit organizations, public or private nonprofit camps, and public or nonprofit private colleges and universities. The federal government provides the funds to each approved organization and sets the rules. Most of these programs have some kind of recreational or educational activity associated with the food program (USDA 2013c).

The Special Supplemental Nutrition Program for Women, Infants, and Children (WIC) was created in 1972 as an amendment to the Child Nutrition Act. It provides healthcare and nutrition to women, infants, and children under the age of five who come from low-income families. The program was the outcome of a White House Conference on Food, Nutrition, and Health organized by the Nixon administration in 1970. Federal funds provide vouchers so that participants can obtain food products at grocery stores including milk, cheese, eggs, whole wheat bread and grains, fish, beans, and peanut butter. The WIC program assists slightly more than half of American infants born today (USDA 2013e).

The Supplemental Nutrition Assistance Program (known more commonly as the Food Stamp program) was created on an experimental basis in 1962, established as an ongoing program by the Food Stamp Act of 1964, and modified by legislation numerous times since then. Today the program provides debit cards to low-income American families to assist them in purchasing food at grocery stores. Over 40 million Americans receive food stamps today. Almost half of the people supported are under age 18. The Roosevelt Administration first experimented with a food stamp program in 1936, which it operated as a regular, ongoing program from 1939 to 1943 (USDA 2013d).

Thus, it is clear that the federal government has found various ways to ensure that millions of needy American children are able to receive nutritious meals. These meals are served in many of the environments in which they are most needed: to infants and children of poor families at home or in day-care centers, breakfasts and lunches to school-age children during the school year, and to school-age children from poor neighborhoods as part of summer enrichment programs. These various programs represent a deep commitment on the part of the American public and its lawmakers to use formal means to ensure that children regularly have nutritious meals available to them.

4.2 USDA Evaluation of School Lunches

For the past 20 years, health advocates have been concerned about the role of school lunches in childhood obesity in America. We consider here first the government and private foundation studies of school lunch programs, which tend to be scientifically based, then turn to private critiques, which tend to be more personal and emotional. Both have contributed to the public understanding and debates on school food.

The first governmental review of the nutritional quality of the meals served in the national school lunch and breakfast programs was conducted by the USDA in 1993, using data collected during the 1991–1992 school year. The study found that, while the meals met the targeted levels for minerals and vitamins, they were not in conformance with the 1989 Dietary Guidelines for Americans (the periodic set of nutritional guidelines produced by the USDA based on the latest scientific knowledge) concerning total fat and saturated fat (Burghardt and Devaney 1993). In response to these findings, the USDA initiated efforts to bring school lunches into conformance with its guidelines. The USDA's 1995 School Meals Initiative for Healthy Children worked both to assist school food service personnel to prepare healthier meals, e.g., offering them computerized nutrient analysis tools and to encourage children to eat these meals (Lohrey 2013). The School Meals Initiative succeeded in bringing lunch guidelines into conformance with the Dietary Guidelines with respect to total fat and saturated fat, and it set goals for the number of calories served in federally subsidized school breakfasts and lunches.

A second USDA School Nutrition Dietary Assessment Study was released in 2001, using data from the 1998–1999 school year (Fox et al. 2001). This study found that—despite progress—the 1995 School Meals Initiative guidelines had still not been achieved, and a new call was made for achieving these goals by 2005. More specifically, between 1991 and 1999 there was statistically significant improvement in the fat content of the lunches served without compromising other nutrients, but the schools were still far from achieving the fat goals for meals actually served (as opposed to meals prepared); in 81 % of elementary schools and 91 % of secondary schools, lunches were offered that conformed with the Dietary Guidelines; however, many students did not select these options but instead chose foods that contained more fat than the guidelines allowed. School breakfast offerings in this period made similar improvements to those made in school lunches.

A third USDA School Nutrition Dietary Assessment Study was released in 2007, based on data from school year 2004 to 2005. It found that 85 % of the schools offered lunches meeting the USDA standard for protein, vitamins A and C, calcium, and iron; and that more than 70 % of the time students selected meals meeting these standards. However, less than a third of the school lunches were in conformance with USDA standards for total fat or saturated fat. Comparing 1998–2004, there was no significant improvement in the percentage of lunches that met the total fat standard, but there was a doubling in the percentage (from 15 to 34 % in elementary schools, 13 to 26 % in secondary schools) of schools meeting the saturated fat standard. Thus, over the period from 1991 to 2004, while there was some progress in meeting USDA nutritional guidelines, much remained to be done.

In 2009 the Institute of Medicine released a report entitled *School Meals: Building Blocks for Healthy Children*, prepared at the request of the USDA (Stallings et al. 2009). The report recommended a new set of nutritional standards based on recent scientific evidence that had caused the Dietary Guidelines for Americans and the Recommended Daily Allowances to be updated. The report noted that the cost of supplying meals, especially breakfasts, in conformance with the new standards was likely to increase because of the need to include larger quantities of fruits, vegetables, and whole-grain foods—all of which were more expensive than ingredients typically used in school meals up until that time. The new standards called for increased amounts of protein, vitamins, and minerals, while decreasing the amount of sodium. The only way to achieve these goals was to increase the amount and variety of fruits and vegetables, replace most refined grain products with whole-grain products, and replace whole and 2 % milk with 1 % or fat-free milk. The report also called for the number of calories in the meal to be capped.

While there was little dispute of the findings in the Institute of Medicine report, there were structural problems in implementing solutions. These recommendations called for changes that do not fit well with either the practices of the large industrial food service companies that supply most American foods or with the budgets that schools have available for food purchases.

4.3 Private Critiques of School Lunches

There have also been criticisms of the school lunch program from private citizens. Three of the most vocal critics have been Jamie Oliver, Sarah Wu, and Avis Richards. (These criticisms are somewhat similar in content to the criticisms of fast food available to the general public, such as Schlosser 2001; Spurlock 2004.) One well-intentioned, but not necessarily helpful advocate for healthy school meals has been the British chef Jamie Oliver. Through his food show on British Channel 4, Oliver convinced the British government to change the standards for school lunches to serve at least one serving of fruit and vegetable at each meal, ban sweetened carbonated beverages, and limit deep-fried food to twice a week (Garner 2009; BBC News 2005). Oliver believed he could have the same salutary effect in the United States, through his American television show *Jamie Oliver's Food Revolution*. He attempted to transform school meals from using highly processed foods to foods prepared from scratch with fresh ingredients. In the first year of his television series, the focus was on Huntingdon, West Virginia, the U.S. city with the highest rate of obesity. The second and final year of the television program focused on the bureaucracy of the Los Angeles school district as an impediment to bringing healthy food to schools. In a follow-up to his show, Oliver circulated a petition calling for better food at school and obtained 400,000 signatures by April 2010. However, he did not receive the same kind of formal policy response in the United States that he had received in Britain, perhaps because of the political might of the industrial food suppliers to the schools, the weak budgets of the local schools, the fact that decisions about school lunches were made at the local rather than the national level in the United States, and resentment over a foreigner coming in and embarrassing U.S. families and school districts.

Another of the critics is Sarah Wu, a schoolteacher and mother of school-age children in Illinois. One day she forgot her lunch and ate the school lunch. Appalled at what she was served, she began to eat the school lunches every day for a year and blog as Mrs. Q about her experience (anonymously, since she was afraid of losing her job if her identity was revealed). Her blog attracted thousands of readers daily, and eventually she revealed her identity and published a book, *Fed Up with Lunch: The School Lunch Project* (2011), which pulled together the material from her blog. (A detailed account of Sarah Wu's efforts can be found in Aspray, Royer, and Ocepek 2013.)

A third criticism came from the documentary filmmaker Avis Gold Richards. Her film, entitled *Lunch*, examines the National School lunch program and its effects on children through interviews with doctors, teachers, farmers, and others (Birds Nest Foundation 2011).

The main purpose of these efforts by public citizens is to make the public aware of the general problem of lack of nutritious foods in the schools and the relationship to childhood obesity. Their efforts change the conversation away from merely fat and salt content to address the quality of meals that children eat. While these three efforts raised public awareness, they did not have much effect on the formal policy process, probably because of the distributed structure of political control over school food in the United States and the political power of the food industry.

4.4 Competitive Foods in the Schools

Food and beverage companies such as fast-food restaurant chains, snack food makers, and soda manufacturers have a deep interest in gaining children as consumers. It is their desire to use the schools as both a place to promote their products and a place to sell them.

Commercial activities in public school go back at least to the nineteenth century. For example, in 1890 a paint company provided a corporate-sponsored handout for art classes on primary and secondary colors. Over the years since the Second World War, commercial activities in the public schools have become increasingly common. Sheila Harty's popular book, *Hucksters in the Classroom* (1979), together with two other studies, has made these commercial activities familiar to the public. The first of these studies was conducted by Consumers Union, the publisher of the highly trusted *Consumer Reports*. In 1993 Consumers Union began to collect data about commercial activity in schools, releasing its final report in 1998 (Consumers Union 1998). Four types of in-school commercialism are examined: sponsored educational materials, contests and incentive programs, Channel One and other media programs that include commercial advertising, and in-school advertising. The report extends far beyond food and nutrition issues, but we restrict the discussion here to this aspect of the report. (For more on commercialism in schools and more generally on the commercialization of children, see Palmer and Sofio 2006; Linn and Golin 2006; Linn and Novosat 2008; Schor 2005a, b; Nadeau 2011; Kelly et al. 2011; John 1999; Kunkel et al. 2004; Hill 2011; Jones et al. 2010; Calvert 2008.)

In its section on educational materials, the Consumers Union report notes that all but one of the 21 educational materials about nutrition that they sampled were sponsored by food companies or food industry associations. One common and troubling pattern noted was that the focus was on nutritional problems that their products did not have, while avoiding discussion of the nutritional limitations of their own products. For example, the educational material from cereal manufacturer Kellogg, *Kids Get Going with Breakfast*, which includes a poster and activity sheets for children as well as an activity guide for parents, focuses only on the problem of fat content and does not mention the nutrition problems associated with high sodium and sugar levels. In some cases, the educational materials are thinly veiled commercialism. In the *Prego Thickness Experiment*, for example, sponsored by the maker of Prego spaghetti sauce (Campbell's), students are taught "scientific thinking" so that they can demonstrate the claim made by Campbell's in its television ads that Prego is thicker than the rival sauce, Ragu. A problem with these one-dimensional commercial efforts at nutritional education is that children do not have a strong understanding of the relationship between foods and nutrition, so they will have difficulty in identifying practical advice about what to eat that will be healthy for them. (See, for example Noble et al. 2000 on the understanding of primary school children in Britain.)

The General Accounting Office published the other report, entitled *Public Education: Commercial Activities in Schools* (U.S. General Accounting Office 2000). It provided a classification of four types of commercialism in school—different from the Consumer Union scheme—sale of products (e.g., sweetened beverages), direct advertising (e.g., ads on vending machines and scoreboards), indirect advertising (e.g., corporate sponsored educational materials or teacher training), and market research (which they found was not yet common). The GAO found that commercial activities in schools were seldom regulated at the federal level and regulated by only 19 states. The report found state laws were often narrow rather than comprehensive and of considerable variation in scope. (Also see Center for Science in the Public Interest 2007 on state food policies.) To the extent there was any local regulation of commercial activities in schools, the report noted, it was provided by school boards or school principals—and these local groups often had a conflict of interest because the commercial activities could bring in needed cash or equipment donations to the publicly underfunded public schools. Even so, the commercial activities did not generate a lot of revenue for the schools—typically between \$3 and \$30 per student for each school year.

Both the Consumers Union and GAO reports discuss Channel One, which is a news program shown to students in grades 6–12 in more than 350,000 classrooms across the country. The company provides and maintains the technology (a satellite dish, wiring, and television monitors for each classroom) free of charge for as long as the school airs its show at least 90 % of school days in at least 80 % of its classrooms. Channel One programming includes 10 min of news and current events programming and 2 min of advertising. Candy, pretzels, gum, and soda are all advertised. Advertisers value this access to the captive student audience and pay an average of $194,000 for a 30-s slot to be run nationwide. Parent and teacher organizations have long opposed Channel One in their schools, but principals and school boards steadfastly support these contracts in order to receive the free equipment. At least three court cases have been filed to try to remove Channel One from particular school districts [*State v. Whittle Communications* (NC, 1991); *Dawson v. East Side Union High School* (CT 1994); *Wallace v. Knox County Board of Education* (1993)], but in each case the court sided with the local authority of the school board to set contracts. Consumers Union found that poor, southern, and ethnic school districts are more likely than wealthier, white school districts to enter into contracts with Channel One. (Also see American Academy of Pediatrics 2006; Graff 2008; Palmer and Sofio 2006.)

The story is similar with Star Broadcasting, which pipes popular music, together with food ads, into school hallways and cafeterias. Another company, ZapMe, installed up to 15 free computers, a printer, and an Internet connection into school computer labs at 12,000 schools in return for the right to place ads on the computers. In 1998 both the Washington, D.C.-based nonprofit Commercial Alert and consumer advocate Ralph Nader spoke out against ZapMe and more generally against commercialism in the schools. ZapMe ran into trouble for collecting private information from the student who used these computers and exited the computers-for-school business in 2000 (Palmer and Sofio 2006; Schiffman 2000).

Even if the schools offer healthy lunches, schools may be contributing to obesity if the students choose to eat less healthy alternatives—either from the *a la carte* menu in the lunchroom or from so-called competitive foods that are offered in the school vending machines, school snack bars, or school stores. For example, in the year 2000, students could buy sweetened drinks at 58 % of elementary schools, 83 % of middle schools, and 94 % of high schools (U.S. National School Health Policies and Programs Study 2000; also see French et al. 2003; Story et al. 2008). Both corporate advertising and access have a bearing on students' preferences to certain kinds of food. (See Chernin 2008; Hoek and Gendall 2006; Cornwell and McAlister 2011; Rosenheck 2008; also Chap. 3.) In 1970 the Child Nutrition Act was extended so that the USDA could regulate competitive foods in schools (Mello et al. 2008). Several studies identify the role of branding, packaging, and cafeteria ambience in eating healthy food choices at school, including lessons learned from the fast-food restaurants (Johnston et al. 2009; also see Bowman et al. 2004; Robinson et al. 2007; Rosenheck 2008; Schlosser 2001; Pereira et al. 2005).

Already by the 1960s, vending machines selling sweetened drinks were common in schools. However, consumption more than doubled before the end of the century. The new, exclusive contracts between schools and soft drink companies for "pouring rights" were a contributing factor (French et al. 2003; Palmer and Sofio 2006; Linn and Novosat 2008; Alderman et al. 2007; Nestle 2000). A drink company would commonly sign a contract with a school district that called for the school district to sell only that beverage company's products in its schools. The contracts could be lucrative if the schools met certain sales quotas, and the schools would

Dear Principal

Here we are in year two of the great Coke contract. I hope your first weeks were successful and that pretty much everything is in place (except staffing, technology, planning time and telephones).

First, the good news: this year's installment from Coke is "in the house" and checks will be cut for you to pick up in my office this week.

Now the not-so-good news: we must sell 70,000 cases of product (including juices, sodas, waters etc.) during the first three years of the contract.

The math on how to achieve this is really quite simple. Last year we had 32,439 students, 3,000 employees and 176 days in the school year. If 35,439 staff and students buy one Coke product every other day for a school year, we will double the required quota.

Here is how we can do it: 1. Allow students to purchase and consume vended products throughout the day. If sodas are not allowed in classes, consider allowing juices, teas and waters. 2. Locate machines where they are accessible to the students all day. Location, location, location is the key. 3. A list of Coke products is enclosed to allow you to select from the entire menu. 4. A calendar of promotional events is enclosed to help you advertise Coke products.

I know this is "just one more thing from downtown," but the long-term benefits are worth it.

Thanks for all your help.

John Bushey - The Coke Dude

Fig. 4.1 Letter to Principals in the Colorado Springs, CO School District. *Source*: Brazier (1999) (Verbatim)

often create an environment in which the soft drink sales would increase. Consider the widely publicized letter sent by the District Executive Director of School Leadership to the principals in the Colorado Springs, CO school district as quoted in entirety in Fig. 4.1.

In 1972 the National Soft Drink Association secured an amendment to the bill reauthorizing the national school lunch program to allow competitive foods in the schools. By the mid-1970s there were frequent complaints from both parents and public health advocates about the presence of vending machines in the schools selling sweetened drinks and snacks. When the Carter administration came into office in 1977, the USDA attempted to amend the National School Lunch Act so as to limit these sales of sweetened drinks and other low-nutrient foods. However, the food industry lobbied vigorously against change. A compromise reached that same year provided the USDA with regulatory authority over competitive foods, but only after the USDA agreed not to ban competitive foods and drinks but only place restrictions on those of low nutritive value. The USDA proposed prohibiting low-nutrient competitive foods until after the last lunch period. They were unable to get this regulation approved in 1977, but they did so in 1979. The USDA regulation limits the sale of "food of minimal nutritional value (FMNV)" to after school times and only in designated areas located at a distance from the school cafeteria (Alderman et al. 2007).

The arguments employed by industry representatives at that time are similar to those heard today: (1) there is no scientific evidence causally linking sweetened foods to dental decay (today it would be obesity); (2) any food can be eaten in moderation, and it is the responsibility of the individual to do so; and (3) any regulation should be left to the local school board, not to the federal government. Consider this comment from James E. Mack, the President and General Counsel of the National Confectioners Association:

> There's no reason why a child should not have a soft drink or a candy bar … School authorities should teach moderation rather than try to prohibit the sale of these items … Tooth decay is caused by a combination of factors … If those advocating curtailment of school sales were as interested in encouraging children to brush their teeth … as they are in trying to take candy away from them, they would accomplish much more.
>
> (James E. Mack, President and General Counsel, National Confectioners Association, quoted in Alderman et al. 2007, fn 51).

When lobbying failed, the National Soft Drink Association sued the USDA (*National Soft Drink Association v. Block*, 1983). The court ruled mostly in the industry's favor: while FMNV could not be sold in the cafeteria at lunchtime in direct competition with the school lunch, that was the only time and only restriction that the USDA could place on the sale of FMNV on school premises—all other times and places were the province of the local school board to regulate. This ruling led to a bonanza for the food industry in the schools. For example, vending machines were relocated to places so that students would have access to them all day—often even during the lunch hour, contrary to the court ruling. Industry also skirted USDA regulations about what constituted food of minimal nutritional value through the creation of new products, such as sports drinks and energy snacks, that offended the spirit if not the letter of the law.

A study by anthropologist Deborah Crooks (2003) presents a clear picture of the dilemma facing teachers and administrators in a public school in the Appalachian region of eastern Kentucky. This is a poor school district, where three-quarters of the students qualify for free or subsidized lunches. These poor schools are the ones that statistically have among the highest percentage of obesity among their students. As Crooks explained, the problem is that the decision about allowing low-nutrient foods into the school pits a worthy health objective against a worthy career and educational objective. The snack room in this school brings in $8,000 a year in profit, which is used by the school to pay for telephone bills, teacher equipment such as slide projectors, and school trips for children who cannot afford them. At other schools across Kentucky these profits are used for "sports equipment, music programs, guest speakers, field trips, student awards and incentives, books, instructional materials, computers, paper supplies, and extra funds to subsidize school lunches" (Crooks 2003). Even though the availability of these low-nutrient foods undermines the message offered in the physical and health education classes and sometimes leads to unruly behavior when the students are hyped up on sugary foods, the school administrators do not believe they could afford to give up this income stream. Some see this income as the only way to keep school interesting enough that students will not drop out, and keeping the students in school is the only way to break the chain of poverty that is passed on from parents to students who do not complete their education. When asked why not sell more nutritious snacks at the school, the principal indicated that candy, chips, and soda generated the greatest income.

Even in public schools in wealthier school districts, the principals are generally in favor of the corporate presence (Alderman et al. 2007). In the 2004–2005 school year, competitive foods were available in 73 % of elementary schools, 97 % of middle schools, and 100 % of high schools (Fox et al. 2009). By the 2009–2010 school year, the numbers had fallen overall so that approximately half of all public school students had access to competitive foods in school (Turner and Chaloupka 2012; also see Hartline-Grafton 2010). Competitive foods were slightly more available in suburban and rural schools than urban schools, and slightly more available in schools in the South (Turner and Chaloupka 2012).

A General Accounting Office (2000) report notes that contracts with food companies provide significant funds to pay for scoreboards, band uniforms, and other discretionary expenses. The schools receive a fixed percentage of the revenue from sweetened drinks sold in their schools under these pouring rights contracts, which make that school the exclusive territory of the particular beverage company. In many cases, the food company also donates equipment or even curricular materials. One school district reported that its beverage contract netted $1.5 million annually.

Under these contracts, the school serves as a marketing canvas for the company. Corporate logos appear on school buildings, playing fields, and the screen savers on the school's computers; ads appear in school newspapers and on school buses; Channel One runs 2 min of advertisements for sodas, candy, and fast food meals during its daily program. There are even food-company sponsored curricula being used in some schools, in nutrition and science classes.

Throughout the remainder of the 1980s and the 1990s, control of competitive foods was left largely in the hands of state and local legislatures, and individual schools and school districts. This is not to say that there were not attempts at national regulation of competitive food. In 1990 the Citizens Commission on School Nutrition, which is a group of doctors, educators, nutritionists, and school food workers organized by the Center for Science in the Public Interest, recommended regulations on both the nutritional content of both school lunches and competitive foods in the schools (Burros 1991). The next year, the American Dietetic Association and the American Food Service Association recommended a ban on all competitive foods in the schools. In 1995 the Center for Science in the Public Interest encouraged the USDA, which controlled the federal school lunch program, to require any competitive foods offered in the schools to meet the USDA guidelines on nutritive foods. In 1998 the Center published *Liquid Candy*: *How Soft Drinks are Harming Americans' Health* (Jacobson 1998 1st ed., 2005, 2nd ed. with updated data).

There were small movements toward regulation in response to these continued calls. In 1999 the USDA placed sweetened drinks on the "eat less" category for children of ages 2–6. The 2000 Dietary Guidelines issued by the USDA recommended reducing the consumption of sweetened drinks so as to reduce sugar consumption. However, all of this amounted to very little progress in reducing the unhealthy consequences of competitive food on children during the 1980s and 1990s. It was not until the great concerns about obesity, epitomized by the Surgeon General's report in 2001, that federal regulation again gained political traction. Even then, it took most of the next decade to come to terms with this issue.

The Institute of Medicine was asked by Congress to work with the Centers for Disease Control to develop guidelines for foods in schools. Their joint report, *Nutrition Standards for Foods in Schools: Leading the Way to Healthier Youth* (Stallings and Yaktine 2007), made several major recommendations: (1) meals falling under the nutritional guidelines of the national breakfast and lunch programs should be the main source of nutrition at school; (2) access to competitive foods should be limited; and (3) if competitive foods are available, they should be limited to healthy choices such as fruits, vegetables, whole grains, and nonfat or low-fat milk and dairy products. These recommendations were in keeping with the 2005 *Dietary Guidelines for Americans*, the comprehensive, science-based set of recommendations prepared jointly by USDA and Health and Human Services. The Institute of Medicine report categorized possible competitive foods into three categories: Tier 1 foods (e.g., individual pieces of fruit, raw vegetables, dehydrated fruits, low-fat and low-salt whole grain crackers, low-fat and low-sugar fruit-flavored yogurt) can be made available to all students at any time of the school day. Tier 2 foods (e.g., low-salt baked potato chips, reduced sugar animal or graham crackers, low-sugar and low-fat ice cream bars, and caffeine-free and calorie-free nonfortified soft drinks) can be made available only to high school students—not to younger students—and only after school. Some items (e.g., cupcakes or cookies with too much salt or sugar, gum, candy, fruit smoothies with added sugar, and regular sodas with sugar or caffeine) do not qualify for either Tier 1 or Tier 2 and should not be served at school.

Thus, by 2007 there was a scientifically based understanding of what needed to be done to bring school lunches and competitive foods under control for the health of America's children. Although the nutrition standards—as represented by the *Dietary Guidelines for Americans*—had changed over time, as early as the mid-1980s there was already a general sense, supported by scientific evidence, of what changes should be made to school lunches and competitive foods. However, finding the political will to make these changes was another matter. During that period of inaction, from about 1985 to 2005, childhood obesity almost *tripled* in the United States.

4.5 Legislative and Other Attempted Remedies

Serious efforts to regulate food in the schools began in the late 1990s, during the final 2 years of the Clinton presidency. The pace of legislative activity picked up in the middle of the next decade, especially during the second term of President George W. Bush. At the federal level, several bills were introduced in the 108th Congress (2003–2005), including among others Senator Edward Kennedy's Prevention of Childhood Obesity Act, Senator Tom Harkins's Healthy Lifestyles and Prevention

American Act (HeLP America Act), and Representative Mary Bono's and Senator Bill Frist's Improved Nutrition and Physical Activity Act (IMPACT). Over the next several sessions of Congress additional bills were introduced, including Senators Tom Harken and Lisa Murkowski's and Representative Lynn Woolsey's Child Nutrition Promotion and School Lunch Protection Act, which tried to apply the latest scientific findings to the definition of food of minimal nutritional value in the national lunch and breakfast programs. These various bills typically called for additional scientific research or supported cradle-to-grave wellness programs (Alderman et al. 2007).

Not surprisingly, there was strong industry opposition to all proposed federal legislation related to children and food. The Center for Consumer Freedom, the National Restaurant Association, the Grocery Manufacturers of America, and the Association of National Advertisers all took stances against federal regulation of school food. None of these bills became law.

Congressional action was partly informed by a report to Congress in 2003 by the General Accounting Office (U.S. General Accounting Office 2003). The report called for reducing or replacing foods of limited nutritional value. It also called for branding foods with a logo and packaging foods in the same way in which they were packaged in fast-food restaurants—to take advantage of the attraction children had to fast food. The report recommended using the monthly school lunch menu as a device to educate students, parents, and teachers about nutrition. The report also suggested partnerships between the schools and local businesses and nonprofit organizations to get the word out about healthy eating.

One law that did pass during these years was the reauthorization of the WIC program. This bill mandated that all school districts receiving federal aid (hence, all of them) implement wellness programs in time for the 2006–2007 school year. Unfortunately, there was no enforcement mechanism and an Institute of Medicine report on nutrition standards for school food noted that the wellness programs in the local schools were of great variation and generally not very successful (Stallings and Yaktine 2007). Many school districts had taken action since 2002 to limit portion sizes, remove fried foods and sweetened drinks from the *a la carte* menu in the school cafeteria, and introduce more fruit and vegetables. But in a 2004 survey of the largest school district in each of the 50 states and the District of Columbia, not a single district had yet met the 2005 Institute of Medicine guidelines on healthy foods in schools (McGinnis et al. 2006; see Institute of Medicine 2007; Koplan et al. 2005 for other relevant studies from the Institute of Medicine; also Greves and Rivara 2006). The Institute of Medicine also reported significant pushback from parents and students who wanted to give the students free choice over their diets, and from school administrators who were not strongly supportive of a healthy food initiative when they were also obligated to spend money and effort to meet the demand to improve academic performance (Mello et al. 2008).

The first state law regulating food sold in the schools was enacted in California in 1979 (Mello et al. 2008). However, it was not until the period from 2000 to 2005 that most states addressed this issue. In 2005 alone, 42 states introduced bills concerning food in schools and more than 20 states passed legislation. A study by the Center for

Science in the Public Interest graded these programs, finding considerable variation in what was regulated and identifying Kentucky and Oregon as having the regulations most likely to be the effective in controlling obesity (Alderman et al. 2007).

Similarly, some cities—or their school districts—began to regulate commercial food activities in the schools during this first decade of the new century. Oakland, California banned all school sales of soda and candy. Los Angeles banned sodas in vending machines. Detroit banned fast-food restaurants within 500 ft of a school. Phoenix banned mobile street vendors within 600 ft of a school during school hours. New York City replaced white with whole wheat bread, required all milk to be fat free, and banned bake sales. Philadelphia enacted a beverage policy that limits drinks sold in the schools to juice, water, and milk in elementary school, and those plus sports drinks in high school (Perdue et al. 2003; Mello et al. 2008).

Beginning in 2004 both the Public Health Advocacy Institute, a legal research center located in Boston focused on public health, and the Center for Science in the Public Interest threatened (but never filed) to sue the food companies. They argued that it is an unfair business practice to sell unhealthy foods to minors in a place (school) where the minors are compelled to be and where they will be tempted. This led the food companies to enter into discussions with these two public interest groups, as well as with the Center for Informed Food Choices, about improved self-regulation of foods in schools (Mello et al. 2008).

The food industry was troubled by this patchwork of regulatory efforts that occurred between 2000 and 2005 for two reasons. First, these regulatory efforts presented a serious public relations problem, implying that the food industry was harming children. Second, with each state and school district establishing different regulations, it was hard for the food manufacturers to produce products that would meet or exceed the requirements of a sufficient number of legislative districts to maintain a sufficiently large potential market for their products.

The first reaction of the food industry was to use its considerable lobbying muscle. In 2002 the National Soft Drink Association ran local campaigns across the country disputing the causal relationship between sugary beverages and obesity, promoting increases in physical exercise of children, and noting the strong economic benefits school districts receive through their relationships with local businesses including the bottlers who supplied the schools.

Lobbying was followed by voluntary self-regulation. In 2003 CocaCola, the largest beverage company, announced guidelines to eliminate soda sales in elementary schools, although these guidelines were voluntary for both the school districts and the local bottling companies. Two years later, the American Beverage Association suggested a voluntary policy to the school districts, the beverage companies, and the local bottling companies—limiting drinks in elementary schools to water and pure juice, removing sugary drinks from middle schools, and restricting sugary drinks in high schools to no more than half the products sold (Mello et al. 2008).

This climate of state and local regulations, threatened lawsuits, and initial attempts at self-regulation, together with a drop in sales within schools of regular soda in favor of water, diet soda, sports drinks, and fruit juices over several years, set the stage for an initiative by former president Bill Clinton. In 2006 the Alliance

- The process by which nutrition standards were determined was not transparent.
- There was no objective input from the scientific community.
- Clear benchmarks (e.g. sugar intake) were not established in advance.
- The agreement was less restrictive than some of the state regulations.
- There is no evaluation by people not funded by industry.
- Some sport and energy drinks do not fall under the guidelines.
- The guidelines are not mandatory. For companies that choose to participate, the beverage companies cannot compel local bottling companies to follow these regulations.
- The agreement does not abrogate existing contracts between school districts and local bottlers, which might remain in effect for as many as ten more years.

Fig. 4.2 Criticisms of the agreement of the bottling industry with the Clinton Foundation and the American Heart Association. *Source*: Sharma et al. (2010) and Mello et al. (2008)

for a Healthier Generation, organized by the William J. Clinton Foundation and the American Heart Association, reached a 3-year agreement with the American Beverage Association to phase out sugary beverages in schools and replace them with healthier alternatives. (On the relation between sugary beverages and weight gain, see Malik et al. 2006). These goals could only be achieved in collaboration with the local school boards, which held the contracts with the bottling companies. The target was, by the 2008–2009 school year, to renegotiate these contracts in 75 % of the school districts that held current contracts with bottling companies, and with 100 % of the school districts a year later. The CocaCola Company, PepsiCo, Dr. Pepper Snapple Group, and their respective bottling companies removed full-calorie soft drinks from schools across the country and replaced them with lower-calorie and age-appropriate serving sizes. Full-calorie soft drinks shipment to schools dropped by 95 % between 2004 and 2009, and beverage calories shipped to schools dropped by 88 %. By 2009, over 98 % of school districts were following the guidelines (American Beverage Association 2010). These changes did not harm the beverage companies financially because they were able to replace the sale of sweetened sodas with sales of other beverages containing fewer calories and sometimes more nutrients. In 2006, the Clinton Foundation reached an agreement with snack manufacturers about nutrition standards for snacks sold in schools, modeled directly upon the sweetened beverage agreement (Mello et al. 2008; Alderman et al. 2007).

The program worked reasonably well in part because of the buy-in of the three largest beverage companies (CocaCola, PepsiCo, and Cadbury Schweppes), which together account for more than 70 % of the American beverage market. Nevertheless, as Fig. 4.2 indicates, the program has had its critics.

Another major achievement was reached with industry cooperation in early 2010. Many of the school breakfasts and lunches served in America are dependent on food service providers who plan menus, provide meals and food materials, and even create the dining spaces within schools. The three largest foodservice companies serving American schools—the US firm Aramark, the French firm Sodexo, and the British firm Chartwells—all agreed to increase the amount of fruit, vegetables, and whole grains; move to low-fat and fat-free milk products; and follow the guidelines about fat, sugar, and sodium in the Institute of Medicine 2009 report on school meals. To give a sense of the scope of this agreement, ARAMARK serves approximately 300 million school meals annually. The decision by these foodservice providers was made directly in response to Michelle Obama's Let's Move initiative for healthier schools. (See Chap. 6 for a discussion of this initiative.)

The most significant piece of federal legislation about food in schools came as a result of the reauthorization of the Child Nutrition Act, which pays for the national breakfast and lunch programs as well as WIC, the Summer Food Service program, and the Child and Adult Care Food Program. In December 2010 President Obama signed the Healthy, Hunger-Free Kids Act, which had passed Congress with broad, bipartisan support. It was the largest reform in the school foods program in 15 years. The Act implements the recommendations in the Institute of Medicine *School Meals* report by giving the Secretary of Agriculture the authority to issue regulations to schools consistent with the report's recommendations. Schools will be required to offer both fruit and vegetables every day, increase whole grains, serve only low-fat and fat-free milk products, limit calorie amounts by age, and limit intake of fats and sodium. The Act provides resources so that schools can obtain fresh produce from local farms and community gardens. The Secretary of Agriculture is also given the authority to set nutritional standards for all other foods offered in school, including those sold in vending machines, school stores, and school snack bars, but not those for special events such as bake sales and parties. In 2013, using this provision in the law, the USDA essentially banned the sale of snack foods anywhere in the schools, including vending machines (Strom 2013). The Act also makes the more than 100,000 students on Medicaid automatically eligible for free lunches. The estimated cost is $4.5 billion over 10 years. School districts are to be audited every 3 years to determine whether they have met the specified federal nutrition standards.

There have been some predictable complaints about this law. Some children do not like the foods or the amounts of food served under the new program. For example, a group of high school students produced a music video entitled "We are Hungry" that claims that the number of calories is insufficient for active teens such as those on the school sports teams (Huffington Post 2012). A report by conservative Fox News claims that students are rejecting the healthy lunches that are being served and that there is a black market for chocolate syrup in the schools, though little evidence is given (Fox News Insider 2012). A military website has expressed concerns about the wastefulness of the program when students throw their vegetables in the trash (Smiley 2013). One conservative commentator has complained about the waste of tax dollars because some schools can enroll children for free school lunches, based on general demographic information, without the parents having to apply for these programs. This was an efficiency effort on the part of the Obama Administration, but it cuts against traditional Conservative policies that want people to prove their eligibility for federal support (Pullman 2013). An editorial from the food service industry claims that the bill is harmful to the food service workers because they do not have the adequate skill set to plan menus, cook from scratch, source fresh foods, and employ best practices of business management

(Smythe 2013). The implementation of the program, affecting approximately 100,000 schools across the 50 states, has gone relatively smoothly for such a massive undertaking—far more smoothly, for example than the implementation of the Obama health care initiative. It is far too early, however, to know the long-term impact these changes will have on the fight against childhood obesity and the improved health of American youth. (For a recent evaluation, see Josel 2013.)

4.6 Conclusions

Many formal approaches have been taken or suggested to reduce the contribution of schools to childhood obesity. These include taxing unhealthy foods available in the schools, decreasing the cost of healthy foods through agricultural subsidies so that healthy alternatives are economically feasible in the schools, making sure there is adequate water available as an alternative to sweetened drinks, requiring vending machine companies to replace traditional snack items with fresh fruit, removing all unhealthy foods from schools, using zoning to restrict the proximity of fast-food restaurants to schools, banning commercial advertising in schools, and improving the image of healthy choices through public service advertising and counteradvertising (Frieden et al. 2010; Jacobson and Brownell 2000).

Other scholars have suggested performance-based regulation of the food industry as the solution (Sugarman and Sandman 2007). Unlike what they call "command and control" strategies of using tort law to sue fast-food restaurant and soft drink companies or use regulatory procedures to order these companies or school officials to behave in a certain way, they propose "performance-based regulation" in which the outcome targets are specified but not the means to achieve them. For example, these companies may be given a target obesity rate to achieve for a certain geographic region but not told how to achieve the result. This strategy is modeled after the No Child Left Behind legislation for education. However, there are numerous critics of that legislation, and there is not yet reason to believe that performance-based regulation will work with childhood obesity.

Industry stymied federal regulation of both advertising (through the television, Internet, and other means) and school food (both federally supported lunches and competitive foods) for more than 30 years. With the public recognition of the obesity epidemic as of 2001 and the increasing body of scientific information about nutrition, there were major efforts during the first decade of the new century. The capstone of these efforts is the Healthy, Hunger-Free Act of 2010. While there has been significant progress, it is too soon to know if these regulations will help enable a long-term solution to the child obesity epidemic.

Industry has called for self-regulation and local regulation of these issues. (For a general discussion of self-regulation, see Federal Trade Commission and Department of Health and Human Services 2006; Mello et al. 2008; Sharma et al. 2010; Whitaker et al. 1997; Simon 2006) Indeed, industry has made some significant progress, such as the reduction of sweetened drinks in the schools and the removal of advertising on the Disney Channel. However, there continue to be many skeptics of the efficacy of self-regulation, and comparative analyses of Canada and Australia support this skepticism. (See, for example, Kent et al. 2012; Wilde 2009).

In an interesting study, three academics (Sharma et al. 2010) compared self-regulation in the food industry with self-regulation that has been historically tried in four other industries. They found that in the cases of forestry and fisheries—where there is a dwindling resource that faces the threat of overuse—self-regulation has worked relatively well. However, in the cases of tobacco and alcohol—where the self-regulation is motivated by external threat of negative public attitude, government regulation of key business practices, and litigation—self-regulation has been a failure. In the case of the tobacco industry for example, young people were encouraged to smoke more under self-regulation. The authors conclude that the food industry is motivated by interests more like those of the alcohol and tobacco industries than the forestry and fisheries industries, and they are pessimistic about the chances of self-regulation in the food industry to lead to public good. (For more information about the relationship between food policy and smoking policy, see Kline et al. 2006; Courtney 2006; also see Fig. 3.2 in Chap. 3).

References

Adamick K (2012) Lunch money: Serving healthy school food in a sick economy. Food Systems Solutions, New York

Alderman J, Smith Jason A, Fried EJ, Daynard RA (2007) Application of law to the childhood obesity epidemic. J of Law Med Ethics 35(1): 90–112

American Academy of Pediatrics (2006) Children, adolescents, and advertising. Policy Statement, Committee on Communications. Pediatr 118(6):2563–2569

American Beverage Association (2010) Beverage industry delivers on commitment to remove regular soft drinks in schools, driving 88 % decline in calories. News release. http://www.ameribev.org/news--media/news-releases--statements/more/183. 8 March 2013, Accessed 19 Dec 2013

BBC News (2005) TV chef welcomes £280m meals plan. http://news.bbc.co.uk/2/hi/uk_news/education/4391695.stm. 30 March 2005, Accessed 7 October 2013

Birds Nest Foundation (2011) Lunch the film. http://lunchthefilm.com/. Accessed 13 Dec 2013

Bowman SA, Gortmaker SL, Ebbeling CB, Pereira MA, Ludwig DS (2004) Effects of fast-food consumption on energy intake and diet quality among children in a national household survey. Pediatr. 113(1):112–118

Brazier C (1999) The great education scandal. New Internationalist. 315:7

Brownell KD, Horgen KB (2004) Food fight. Contemporary Books, Chicago

Burghardt J, Devaney B (1993) The school nutrition dietary assessment study: Summary of findings. Mathematica Policy Research, Inc, Princeton

Burros M (1991) Eating well, New York Times. http://www.nytimes.com/1991/01/30/garden/eating-well.html?pagewanted=all&src=pm. 30 Jan 1991, Accessed 19 Dec 2013

Calvert SL (2008) Children as consumers: Advertising and marketing. Future Child 18:205–234

Carmona RH (2003) The obesity crisis in America. Testimony before the Subcommittee on Education Reform, Committee on Education and the Workforce, U.S. House of Representatives. 16 July 2003

Center for Science in the Public Interest (2003) Pestering parents: How food companies market obesity to children. Center for Science in the Public Interest, Washington DC

Center for Science in the Public Interest (2007) State school foods report card: A state-by-state evaluation of policies for foods and beverages sold through vending machines, school stores, a la carte, and other venues outside of school meals. http://www.cspinet.org/2007schoolreport.pdf. Accessed 19 Dec 2013. Center for Science in the Public Interest, Washington DC

Chernin A (2008) The effects of food marketing on children's preferences: Testing the moderating roles of age and gender. Ann Am Acad Polit Soc Sci 615:102–118

Consumers Union (1998) Captive kids: A report on commercial pressures on kids at school. Consumers Union, Yonkers

Cooper A, Holmes, Lisa (2007) Lunch lessons: Changing the way we feed our children. Morrow, New York

Cornwell TB, McAlister AR (2011) Alternative thinking about starting points of obesity: Development of child taste preferences. Appetite 56:428–439

Courtney B (2006) Is obesity really the next tobacco? Lessons learned from tobacco for obesity litigation. Ann Health Law 15:61–64

Critser G (2004) Fat land: How Americans became the fattest people in the world. Houghton Mifflin, Boston

Crooks DL (2003) Trading nutrition for education: Nutritional status and the sale of snack foods in an Eastern Kentucky school. Med Anthr Q 17(2):182–199

Cummings RO (1940) The American and his food. University of Chicago Press. Chicago

Federal Trade Commission and Department of Health and Human Services (2006) Perspectives on marketing, self-regulation, and childhood obesity. Federal Trade Commission and Department of Health and Human Services, Washington DC

Fox MK, Crepinsek MK, Connor P, Battaglia M (2001) National dietary assessment study—II: Summary of findings. http://www.fns.usda.gov/sites/default/files/SNDAIIfind.pdf. Accessed 5 Dec 2013. USDA, Washington DC

Fox MK, Gordon A, Nogales R, Wilson A (2009) Availability and consumption of competitive foods in US public schools. J Am Diet Assoc 109:S57–S66

Fox News Insider (2012) Students choose to go hungry rather than eat healthy school lunches. http://foxnewsinsider.com/2012/09/26/students-hungry-eat-healthy-school-lunches/. 26 Sept 2012, Accessed 5 Dec 2013

French SA, Story M, Fulkerson JA, Gerlach AF (2003) Food environment in secondary schools: A la carte, vending machines, and food policies and practices. Am J Public Health 93(7): 1161–1167

Frieden TR, Dietz W, Collins J (2010) Reducing childhood obesity through policy change: Acting now to prevent obesity. Health Aff 29(3):357–363

Garner R (2009) Jamie's school dinners for all. The Independent. http://www.independent.co.uk/news/education/education-news/jamies-school-dinners-for-all-1779635.html. 31 Aug 2009, Accessed 7 Oct 2013

Graff SK (2008) First amendment implications of restricting food and beverage marketing in schools. Ann Am Acad Polit Soc Sci 615:158–177

Greves HM, Rivara FP (2006) Report card on school snack food policies among the United States' largest school districts in 2004–2005: Room for improvement. Int J of Behav Nutr Phys Act 3:1. doi:10.1186/1479-5868-3-1

Hartline-Grafton H (2010) How competitive foods in schools impact student health, school meal programs, and student from low-income families. Food Res Action Cent 5:1–9

Harty S (1979) Hucksters in the classroom: A review of industry propaganda in schools. Center for Study of Responsive Law, Washington DC

Hill JA (2011) Endangered childhoods: How consumerism is impacting child and youth identity. Media Cult Soc 33:347–362

Hoek J, Gendall P (2006) Advertising and obesity: A behavioral perspective. J Health Commun 11:409–423

Huffington Post (2012) Michelle Obama's low-calorie school lunches slammed by "hungry" high schoolers (Video). Huffington Post Blackvoices. http://www.huffingtonpost.com/2012/09/26/

michelle-obamas-low-calorie-school-lunch-video_n_1914394.html. 2 Oct 2012, Accessed 5 Dec 2013
Institute of Medicine (2007) Progress in preventing childhood obesity: How do we measure up? Committee on Progress in Preventing Childhood Obesity. National Academies Press, Washington DC
Institute of Medicine (2009) School meals: Building blocks for healthy children. Food and Nutrition Board. National Academies Press, Washington DC
Jackson DZ (2001) The other epidemic: Deadly obesity. Chicago Tribune. http://articles.chicagotribune.com/2001-12-19/news/0112190065_1_soda-obesity-surgeon-general-david-satcher. 19 Dec 2001, Accessed 18 Sept 2013
Jacobson MF, Brownell KD (2000) Small taxes on soft drinks and snack foods to promote health. Am J Public Health 90:854–857
Jacobson MF (1998) Liquid candy: How soft drinks are harming American's health. Center for Science in the Public Interest, Washington DC
Jacobson MF (2005) Liquid candy: How soft drinks are harming American's health, 2nd edn. Center for Science in the Public Interest. Washington DC
John DR (1999) Consumer socialization of children: A retrospective look at twenty-five years of research. J of Consum Res 26(3):183–213
Johnston Y, Denniston R, Morgan M, Bordeau M (2009) Rock on café: Achieving sustainable systems changes in school lunch programs. Health Promot Pract 10(2):100S–108S
Jones SC, Mannino N, Green J (2010) "Like me, want me, buy me, eat me": Relationship-building marketing communications in children's magazines. Public Health Nutr 13(12):2111–2118
Josel C (2013) How Obama's Healthy, Hunger-Free Kids Act of 2010 is fairing in 2013. Montgomery County Wise Parenting Examiner. http://www.examiner.com/article/how-obama-s-healthy-hunger-free-kids-act-of-2010-is-faring-2013. 6 Nov 2013, Accessed 19 Dec 2013
Kalfa A (2011) Lunch wars: How to start a school food revolution and win the battle for our children's health. Tarcher, New York
Kelly B, Baur LA, Bauman AE, King L, Chapman K, Smith BJ (2011) "Food company sponsors are kind, generous and cool": (Mis) conceptions of junior sports players. Int J of Behav Nutr Phys Act 8:95–101
Kent MP, Dubois L, Wanless A (2012) A nutritional comparison of foods and beverages marketed to children in two advertising policy environments. Obes 20(9):1829–1837
Kline R, Graff S, Zellers L, Ashe M (2006) Beyond advertising controls: Influencing junk-food marketing and consumption with policy innovations developed in tobacco control. Loyola Los Angel Law Rev 39:603–646
Koplan JP, Liverman CT, Kraak VI, (eds) (2005) Preventing childhood obesity: Health in the balance. National Academies Press, Washington DC
Kunkel D, Wilcox B, Cantor J, Palmer E, Linn S, Dowrick P (2004) Report of the APA task force on advertising and children. American Psychological Association, Washington DC
Levine S (2010) School lunch politics. Princeton University Press, Princeton
Linn S, Golin Josh (2006) Beyond commercials: How food marketers target children. Loyola Los Angel Law Rev 39:13–32
Linn S, Novosat CL (2008) Calories for sale: Food marketing to children in the twenty-first century. Ann Am Acad Polit Soc Sci 615:133–154
Lohrey J (2013) School meals initiative for healthy children. Global Post. http://everydaylife.globalpost.com/school-meals-initiative-healthy-children-24923.html. Accessed 5 Dec 2013
Malik VS, Schultze MB, Hu FB (2006) Intake of sugar-sweetened beverages and weight gain: A systematic review. Am J Clin Nutr 84:274–288
McGinnis JM, Gootman JA, Kraak VI, (eds) (2006) Food marketing to children and youth: Threat or opportunity? National Academies Press, Washington DC
Mello MM, Pomeranz J, Moran P (2008) The interplay of public health law and industry self-regulation: The case of sugar-sweetened beverage sales in schools. Am J Public Health 98(4):595–604
Milne AA (1926) Winnie-the-Pooh. E.P. Dutton & Co., New York

Nadeau MÈ (2011) Food advertising directed at children. Quebec Coalition on Weight-Related Problems, Report. http://www.childhoodobesityfoundation.ca/files/files/AMorel_food_advertising_directed_at_children.pdf. Accessed 19 Dec 2013

Nestle M (2000) Soft drink "pouring rights": Marketing empty calories. Public Health Rep 115:308–319

Nestle M (2002) Food politics: How the food industry influences nutrition and health. University of California Press, Berkeley

Noble C, Corney M, Eves A, Kipps M, Lumbers M (2000) Food choice and school meals: Primary schoolchildren's perceptions of the healthiness of foods and the nutritional implications of food choices. Hosp Manag 19:413–432

Palmer EL, Sofio L (2006) Food and beverage marketing to children in school. Loyola Los Angel Law Rev 39:33–77

Perdue WC, Gostin LO, Stone LA (2003) Public health and the built environment: Historical, empirical, and theoretical foundations for an expanded role. J of Law Med Ethics 31:557–566

Pereira MA, Kartashov AI, Ebbeling CB, Van Horn L, Slattery ML, Jacobs DR Jr, Ludwig DS (2005) Fast-food habits, weight gain, and insulin resistance (the CARDIA study): 15-Year prospective analysis. Lancet 365:36–42

Popendieck J (2010) Free for all: Fixing school food in America. University of California Press, Berkeley

Pullman J (2013) School lunch expansion: Another wasteful, unaffordable entitlement. http://townhall.com/columnists/joypullman/2013/11/05/school-lunch-expansion-another-wasteful-unaffordable-entitlement-n1736905/page/full. 5 Nov 2013, Accessed 5 Dec 2013

Robinson TN, Borzekowski DLG, Matheson DM, Kraemer HC (2007) Effects of fast food branding on young children's taste preferences. Arch Pediatr Adolesc Med 161:792–797

Rosenheck R (2008) Fast food consumption and increased caloric intake: A systematic review of a trajectory towards weight gain and obesity risk. Obes Rev 9:535–547

Schiffman B (2000) ZapMe kills computers in the classroom. Forbes.com. November 28. http://www.forbes.com/2000/11/28/1127zapme.html. 28 Nov 2000, Accessed 19 Dec 2013

Schlosser E (2001) Fast food nation: The dark side of the all-American meal. Houghton Mifflin, Boston

Schor J (2005a) When childhood gets commercialized, can children be protected? Paper prepared for the Yale Legal Theory Workshop. http://www.yale.edu/law/leo/052005/papers/schor.pdf. Accessed 19 Dec 2013

Schor J (2005b) Born to buy: The commercialized child and the new consumer culture. Scribner, New York

Sharma LL, Teret SP, Brownell KD (2010) The food industry and self-regulation: Standards to promote success and to avoid public health failures. Am J Public Health 100(2):240–246

Simon M (2006) Can food companies be trusted to self-regulate: An analysis of corporate lobbying and deception to undermine children's health. Loyola Los Angel Law Rev 39:169–236

Smiley S (2013) Healthy hunger-free kids act sends food to the trash. Military Money. www.incharge.org/military-money/story/hungry-free-kids-act-sends-food-to-the-trash. 3 Dec 2013, Accessed 5 Dec 2013

Smythe M (2013) The Healthy, Hunger-Free Kids Act and its effect on school foodservice…so far. Foodservice Equipment & Supplies. http://fesmag.com/features/foodservice-perspectives/10448-the-healthy,-hunger-free-kids-act-and-its-effect-on-school-foodservice…so-far. 2 Jan 2013, Accessed 5 Dec 2013

Spurlock Morgan (dir) (2004) Super Size Me. Distributor: Samuel Goldwyn Films Roadside Attractions

Stallings VA, Suitor CW, Taylor CL (eds) (2009) School meals: Building blocks for healthy children. Food and Nutrition Board, Institute of Medicine. http://books.nap.edu/openbook.php?record_id=12751. Accessed 20 Nov 2013. National Academies Press, Washington DC

Stallings VA, Yaktine AL (eds) (2007) Nutrition standards for foods in schools: Leading the way toward healthier youth. Institute of Medicine Report. National Academies Press, Washington DC

Story M, Kaphingst KM, Robinson-O'Brien R, Glantz Karen (2008) Creating healthy food and eating environments: Policy and environmental approaches. Am Rev Public Health 29:253–272

Strom S (2013) U.S. standards for school snacks move beyond cafeteria to fight obesity. New York Times. http://www.nytimes.com/2013/06/28/business/us-takes-aim-on-snacks-offered-for-sale-in-schools.html. 27 June 2013, Accessed 19 Dec 2013

Sugarman SD, Sandman N (2007) Fighting childhood obesity through performance-based regulation of the food industry. Duke Law J 56:1403–1490

Turner LR, Chaloupka FJ (2012) Student access to competitive foods in elementary schools. Arch Pediatr Adolesc Med 166(2):164–169

U.S. Department of Health and Human Services (2001) The Surgeon General's call to action to prevent and decrease overweight and obesity. http://www.surgeongeneral.gov/library/calls/obesity/index.html. Accessed 18 Sept 2013.

USDA (2013a) Child and adult care food program (CACFP). http://www.fns.usda.gov/cacfp/child-and-adult-care-food-program. Accessed 13 Dec 2013

USDA (2013b) School breakfast program (SBP). http://www.fns.usda.gov/sbp/school-breakfast-program. Accessed 13 Dec 2013

USDA (2013c) Summer food service program (SFSP). http://www.fns.usda.gov/sfsp/summer-food-service-program-sfsp. Accessed 13 Dec 2013

USDA (2013d) Supplemental nutrition assistance program (SNAP). http://www.fns.usda.gov/snap/supplemental-nutrition-assistance-program-snap. Accessed 13 Dec 2013

USDA (2013e) Women, infants, and children (WIC). http://www.fns.usda.gov/wic/women-infants-and-children-wic. Accessed 13 Dec 2013

U.S. General Accounting Office (2000) Public education: Commercial activities in schools. Report No. GAO/HEHS-00-156

U.S. General Accounting Office (2003) School lunch program: Efforts needed to improve nutrition and encourage healthy eating. Report No. GAO-03-506. http://www.gao.gov/new.items/d03506.pdf. Accessed 19 Dec 2013, Washington DC

U.S. National School Health Policies and Programs Study (2000) Centers for Disease Control.

Whitaker RC, Wright JA, Pepe MS, Seidel KD, Dietz WH (1997) Predicting obesity in young adulthood from childhood and parental obesity. N Engl J Med 37:869–873

Wilde P (2009) Self-regulation and the response to concerns about food and beverage marketing to children in the United States. Nutr Rev 67(3):155–166

Wu S (2011) Fed up with lunch: The school lunch project; How one anonymous teacher revealed the truth about school lunches—and how we can change them! Chronicle Books, San Francisco

Chapter 5
Food Policy During the Depression and the Second World War: FDR's New Deal Legislation and Eleanor Roosevelt's Bully Pulpit

> *No one who ever saw Eleanor Roosevelt sit down facing her husband, and, holding his eye firmly, say to him, 'Franklin, I think you should ...' or, 'Franklin, surely you will not ...' will ever forget the experience. ... It would be impossible to say how often and to what extent American governmental processes have been turned in new directions because of her determination.*
>
> (Tugwell 1963)

Many Americans today are aware of the efforts today of Michelle Obama to use the office of the First Lady to improve food, nutrition, and physical education in the United States. These include, for example, the demonstration garden on the White House lawn and the Let's Move campaign. What is less well known is that Eleanor Roosevelt provides a striking parallel to Michelle Obama in having also used her office as First Lady to promote better food and nutrition. In fact, Eleanor Roosevelt pioneered the role of an activist First Lady, making it easier for First Ladies who followed her to engage in these kinds of high-profile activities (Winfield 1990).

This chapter examines formal and informal means to address American food, health, nutrition, and physical fitness during the Depression of the 1930s and the war years of the first half of the 1940s. Franklin Delano Roosevelt was President of the United States for most of this period—from January 1933 until his death in April 1945, about 4 months before the war ended. Thus, the formal approach to food policy during these years is primarily the story of New Deal policies of the Roosevelt Administration.

While many individuals and private organizations contributed to informal food policy in the United States during these years, this chapter will focus primarily on one individual, First Lady Eleanor Roosevelt, and the various ways in which she used the bully pulpit to promote better food, nutrition, and physical exercise among the American people. In particular, we will show Eleanor's role in: (1) *influencing legislation*, including relief from hunger during the Depression through the federal distribution of surplus farm goods; (2) *encouraging federal agencies* to expand their

W. Aspray et al., *Formal and Informal Approaches to Food Policy*, SpringerBriefs in Food, Health, and Nutrition, DOI 10.1007/978-3-319-04966-3_5,

food and health programs, notably concerning the model community in Arthurdale, WV; (3) *changing individual habits* of American citizens, for example through her writings directed at American mothers on diet and physical exercise; (4) *providing coping strategies* during times of stress, including sample budget-stretching meals during the Depression; and (5) *offering reassurance* to individual Americans, such as her wartime messages about rationing. Eleanor's efforts will be set in a framework that describes the formal policies of the New Deal.

Eleanor Roosevelt was at once both a major promoter of her husband's New Deal programs, using the media effectively for these purposes, and a catalyst for additional change, sometimes at odds with her husband and his administration. She embodied American ideals about food that had been promoted in the 1920s by home economists, nutritionists, vitamin advocates, and food processors and that became prevalent in the mid-twentieth century—as described so well by Harvey Levenstein in his study, *Revolution at the Table* (Levenstein 2003). This viewpoint favored morality over pleasure and scientific approach over taste. It resulted in a reduction in the amount of food consumed by Americans, promotion of nutrition and vitamins in particular, and change in dietary habits away from heavy consumption of beef and toward consumption of more fruits and vegetables.

This chapter is structured by interleaving the informal approaches of Eleanor Roosevelt with the formal approaches of her husband's administration. The first section provides background material about Eleanor Roosevelt and her tools for using the bully pulpit effectively. The next section covers the prewar years, the period from the Roosevelt inauguration in 1933 to just prior to the Pearl Harbor attack in December 1941. Depression-era food policies discussed in this section include surplus food; food and nutrition education; employment, diet, and exercise; Social Security; food stamps and school lunches; planned communities; and model meals. The final section covers the war years. Wartime food policies discussed in this section include the use of kitchen appliances as war materiel, nutritional standards and habits, physical education, rationing and price controls, Victory Gardens, and Eleanor Clubs.

5.1 Eleanor Roosevelt's Use of the Bully Pulpit

Eleanor Roosevelt was arguably the most famous and most influential woman in America during the 1930s and 1940s. In 1939 she had a higher popularity rating than her husband, and she generated more news coverage than any other woman in the twentieth century. She regularly reached into the homes of millions of Americans and served as a role model for American women and girls.

In 1920, women received the right to vote with the passage of the 19th Amendment. Throughout the 1920s, a number of women, Eleanor Roosevelt among them, worked to build on this victory by giving women an important voice in politics (O'Farrell 2010). Beginning in 1920, Eleanor sought personal and financial independence and began to move away from traditional women's work such as charity boards and ladies luncheons to activities that advanced Progressive political ideas such as The World Court, equal rights for women in the workplace, and child labor laws. When her husband was partially paralyzed by poliomyelitis in 1921, she convinced him to remain active in politics and began to more actively enable his political career. In her own right, she became a powerful member of the Democratic Party.

After the 1928 election, when FDR became governor of the state of New York, Eleanor relinquished her position as a journalist for the *Women's Democratic News* but replaced this activity by writing articles for mass circulation periodicals. Many of these articles had to do with issues of the domestic sphere, such as family, housekeeping, and education. Eleanor was not particularly interested in the domestic arts, other than to make sure that families had adequate food, clothing, and shelter to live healthy lives; she was more interested in liberal politics, worker's rights, and the exercise of women's political might. But there were expectations in the American public about the appropriate role for the wife of a high-profile politician, especially after the public criticism of Edith Wilson for having taking too active a role in political decision-making at the end of her husband's presidency, when he was ill. So Eleanor had to pursue her goals within the constraints of public expectation.

Upon FDR's move to the White House in 1933, Eleanor struggled to find a role acceptable to her in the administration. One role she did take on was to be the unofficial eyes and ears of the president. During her first 8 years in the White House, through the worst years of the Depression until the beginning of U.S. involvement in the Second World War, Eleanor traveled more than 300,000 miles. Much of this travel was spent observing life of those downtrodden—in Indian reservations, migrant camps, factories, coal mining towns, poor neighborhoods, civil rights conferences, and federal relief projects. She was able to go places and say things that FDR could not, not only because of her husband's physical limitations, but also because she could engage people who the president did not want to officially recognize and address topics that were objectionable to the president's political base. Eleanor gave the president access to a much larger slice of American life than it was possible for him to experience first-hand, but Eleanor also did this in part to advance her own personal interests. She often pushed the boundaries of social welfare beyond what was covered by the New Deal legislation or was comfortable to her husband to act upon.

Eleanor wanted to be more than the White House hostess, but many of the activities she had carried out previously were unavailable to her from a political perspective (Black 1996; Hickok 1962). She did organize the White House conference on Emergency Needs of Women in 1933 and Camps for Unemployed Women in 1934, two topics that were of particular interest to her; she traveled extensively on behalf of the president—as she had since 1921. Throughout the White House years, Eleanor used the bully pulpit to talk about food, family, and other issues related to the New Deal. She used every medium available to her. She presented radio broadcasts (the latest mass media technology), gave thousands of public speeches, served as a magazine editor, and wrote hundreds of articles for general and women's magazines such as *Woman's Home Companion* and *Ladies' Home Journal*. Her syndicated column "If You Ask Me" appeared monthly in *Ladies' Home Journal* and her syndicated newspaper column, "My Day," appeared 6 days a week in newspapers across the country, from 1936 until just before her death in 1962. She published a dozen books, including several while in the White House, including *This is My Story* (1937).

Eleanor made herself available to reporters in a way that is unmatched by any other First Lady. She gave 348 press conferences while in White House—almost 1 a week—the first one occurring only 2 days after the president's inauguration. She limited attendance at her press conferences to women journalists—causing some of the major news networks to hire women so they had someone to cover the conferences. Before the Second World War, the majority of her press coverage addressed what were typically considered to be women's issues, such as food served or clothes worn. In one amusing press conference in 1939 she defended serving hot dogs to the king and queen of England during a picnic on the grounds of her private residence in Hyde Park, New York. The choices of these topics often originated with the questions asked by the press. Eleanor herself had little interest in fashion or taste. Her interest was in clothing that was practical and durable, rather than fashionable, and in food that was nutritious without much consideration for how it tasted. She was much more concerned with the conditions of the poor and working classes than she was with high couture. In this regard she was diametrically opposed to the interests of First Lady Jacqueline Kennedy (Beasley 1983, 1987; Denker 2003; Roosevelt 1958; Ocepek et al. Forthcoming).

Although she offered stories about the White House social life, there was another dimension to her that went well beyond traditional women's issues. She was not hesitant to share her personal opinions, even when they were controversial and sometimes not in accord with those of the president. For example, she spoke out in support of unpopular civil rights actions, and in what came to be seen as a signal event she resigned in 1939 from the Daughters of the American Revolution when they refused to allow Marian Andersen to sing in their Constitution Hall (Beasley 2010; also see the exhibit website for An American Original: Eleanor Roosevelt, National Archives, http://www.archives.gov/exhibits/american_originals/eleanor.html, accessed 30 August 2013.) She presented her speeches without White House clearance. She joined a labor union (the Newspaper Guild) at a time when unions were politically suspect. She lobbied successfully for the Wagner–Steagall Act of 1937, which provided federal subsidies for low-income housing, and unsuccessfully for the Gavagan–Wagner–Van Nuys antilynching bill that same year. There were only a few topics that she would not touch because they would be toxic to her husband's presidency. One example is her refusal to speak out in favor of birth control, which she favored but which would have alienated her husband's Catholic constituency.

Eleanor courted the press, for example, by letting the journalists who covered her press conferences bring their children to the White House to play with the Roosevelt grandchildren and by sending flowers any time one of these reporters became ill. The press supported her by both teaching her to be a better communicator and not pushing hard on controversial questions during the press conferences.

Throughout her years in the White House, Eleanor carefully crafted an image of herself as a wife and mother. She was careful to avoid being seen as elite, letting readers know about troubles in her own early family life such as alcoholism, family problems, child abuse, and abandonment (Dennis 1995). This made her a more empathetic figure, and she was seen as having experienced a life more like what the general public experienced, not someone sheltered from the real world. This enabled her to be effective in communicating to both stay-at-home and working women.

5.2 Formal and Informal Policy Approaches to Food and Nutrition During the 1930s

This section of the chapter concerns food policy during the Depression years of the 1930s. Each subsection covers a different aspect of food policy. The informal efforts of Eleanor Roosevelt are placed in juxtaposition with the formal policies of the Roosevelt Administration.

Before turning to specific food policy areas, consider the general policy environment for food in the 1930s. President Roosevelt's political power base consisted of an alliance of farmers, labor unions, and middle-class consumers. This alliance was a source of tension throughout all three terms of the FDR presidency. While farmers wanted higher commodity prices for farm goods, and while labor unions called for stable or rising wages, middle-class consumers sought to keep costs as low as possible for both food and manufactured goods. Buying power during the Depression dropped because salaries fell more rapidly than food and other prices. By 1935, purchasing power was a major national political issue. The federal government was not big enough at this time to spend its way out of the Depression, and there was an aversion to incurring federal debt to resolve the economic crisis. So the strategy of the New Deal administrators was to use the formal and informal means available to them to encourage individual citizens and private companies to act on behalf of the nation (Jacobs 2005; Cohen 2003).

The farmers had a difficult time during the Depression years. Unlike most of the rest of the country, the 1920s had been a difficult time for farmers because high debt and overproduction had caused a drop in farm commodity prices. In the early 1930s farm prices continued to drop—about twice as rapidly as manufacturing prices. One of the first pieces of New Deal legislation (typical of the formal approach to food policy) was the Agricultural Adjustment Act of 1933. This law taxed food manufacturers in order to provide the funding for subsidies to farmers, who were paid to keep some acreage out of production and to kill off some livestock—as a means to eliminate overproduction and thus raise farm commodity prices. President Roosevelt argued to the public that this law was good for the entire nation, not only for the farmers, because it would provide farm families with sufficient income to be active consumers of manufactured goods (Jacobs 2005).

Because of both bad weather and economic conditions during the 1930s, food shortages and high prices for food led to a number of civil actions. In 1933 when there was public outcry about rising bread and milk prices, President Roosevelt tried to blame the problem not on the farmers but on the greedy middlemen and on monopolistic practices in the food manufacturing industry. The drought of 1934 led to high meat prices the following year, which caused protests by housewives in New York, Minneapolis, Chicago, and Detroit and boycotts against butchers. In the summer of 1936 there were again public protests over the cost of meat, and in 1937 the rising cost of bread was prominently mentioned in the national press. Milk boycotts occurred. These protests were part of a growing consumer movement that was characterized by the rise of cooperatives to help with family needs such as food and medical care, the creation in 1936 of Consumers Union and its publication of *Consumer Reports*, and popularity throughout the decade of books such as *Your Money's Worth* (Chase and Schlink 1927) and *100,000,000 Guinea Pigs* (Kallett and Schlink 1933). Women were highly active in the consumer's movement through such organizations as The American Association of University Women, The National League of Women Voters, the Young Women's Christian Association, and The National League of Women Shoppers. This gave Eleanor Roosevelt a ready-made audience for her remarks on protecting consumers and in support of consumerism as a patriotic act. Typical of the formal approach, the federal government took antitrust action against the milk industry in Chicago in 1938. (For the historical context, see Mayer 1989; Jackson 1968; McGovern 2006; Cohen 2003; Jacobs 2005.)

5.2.1 Surplus Food

With sustained high levels of unemployment during the decade-long Depression of the 1930s, American policymakers not surprisingly had serious concerns about the nation's health and nutrition. Several of the New Deal programs of the Roosevelt Administration addressed specific issues of food and nutrition. As the Depression set in, food prices dropped because there was not enough demand from people who could afford to pay. When food prices in the city fell below the cost of packing and transporting the food from the farm, surpluses built up on the farms. Generally, the food was left out on the ground to rot—although this was politically chancy when there were so many hungry people. In 1933 the Agricultural Adjustment Administration (AAA) was established to help out the livelihood of farmers by reducing the glut in the food supply. The AAA regulated the amount of food that could be shipped (e.g., grapes from California and celery from Florida), placed restrictions on the quality of food that could be shipped (e.g., only the higher grades of Texas citrus fruit were eligible for shipment), and established minimum pricing (e.g., in the dairy industry) (Cummings 1940; "Tugwell, Rexford Guy" in Beasley et al. 2000; Civitello 2011; Perrett 1985).

In 1933 the Federal Surplus Relief Administration was created to purchase surplus foods and distribute them to state and local relief agencies. Two years later the organization was renamed the Federal Surplus Commodities Corporation and continued in operation until 1942, when surpluses were greatly reduced by the food demands of the military and the shortage of farm workers because of military conscription. The program had some startup problems: there was not a good distribution system in place and some of the perishable items, especially milk, spoiled in the early days of the program. In another early misstep, to avoid a surplus of pigs on the market, the FSRA purchased piglets from farmers nationwide. However, the slaughterhouses were set up for processing full-grown pigs, and a number of piglets escaped onto the streets of Chicago, causing a public outcry (Elias 2009). State governments were never happy with the federal surplus food distribution program, for they would have preferred to receive the cash equivalent instead of the surplus food. However, the food made an enormous difference for many families on relief.

In her first years in the White House, Eleanor Roosevelt used her press conferences to support proposed New Deal policies, in particular to encourage both federal farm subsidies and food surplus programs for the needy. At her October 5, 1933 press conference, for example, she argued that the surplus food being produced by American farmers should not be destroyed but instead should be given to needy families (Beasley 1983). Later that year, Congress changed its policy and used the surplus food for relief families in America and poor children overseas.

At her April 6, 1934 press conference, Eleanor discussed the results of a survey of milk consumption in 50 American cities that had been conducted by the Department of Agriculture (Beasley 1983). The reason for this survey was a belief that preschool and school children were not consuming enough milk for good nutrition. Eleanor took the opportunity of this press conference to air her view that the economics of milk were out of kilter because of the actions of the middlemen—that the farmers received too little income and the consumers paid too much for milk. The survey data was used as the basis for a milk subsidy program enacted by Congress. It is difficult to know how much Eleanor's support of these early programs mattered to their passage, but she did use her public position to support the legislation.

5.2.2 *Food and Nutrition Education*

There were both formal and informal approaches to food education during the 1930s. There is a story—perhaps apocryphal—of grapefruit distributed through the surplus food program to a Nebraska community where grapefruit had never been seen before. The families who received them complained that the grapefruits were tough even after being boiled for more than 2 h! The Bureau of Home Economics, a division of the Department of Agriculture that had been formed in the 1920s, took the lead in a federal food education effort. It offered educational materials, cookbooks, buying guides, and radio broadcasts to teach American families how to eat nutritiously in a time of want. The Consumers' Council of the AAA taught people about how to get enough vitamins and minerals in their diet (Cummings 1940). But surveys showed that by 1940 there was still a lack of basic knowledge among both high- and low-income American families concerning the basics of vitamins and nutrition (Parran 1940). Forty percent of men reporting for military service were being turned away for physical limitations, and the reason in one-third of the cases was nutritional deficiency (Cummings 1940; Perrett 1985).

In response, President Roosevelt convened the National Nutrition Conference for Defense in 1941, 6 months prior to the attack on Pearl Harbor. The 900 attendees included academics (chemists, biologists, social scientists), health workers, and representatives from government and social service agencies, women's clubs, state nutrition committees, and worker and consumer groups. There were even representatives from the Boy Scouts of America and the YWCA. Much of the discussion concerned daily nutritional standards and how consumption of various kinds of mainstream and ethnic foods would meet those standards, as well as means for distribution of nutritious foods. One important thread through the discussions was nutrition education to the American public. The plan was to use existing organizations for public nutritional education: libraries, parent–teacher groups, church groups, garden clubs, and the like (Cummings 1940).

The USDA had been supporting nutrition research since 1883 and had prepared its first dietary recommendations during the First World War—a 14-page booklet entitled *How to Select Foods* (Nestle 2007; Elias 2009). During the Second World War, it published the *National Wartime Nutrition Guide* in 1943.

In her book, *It's Up to the Women* (1933), Eleanor included chapters on healthy eating and physical exercise. She recommended eating in moderation, with a balanced diet containing appropriate fats, proteins, and vitamins. She offered a week's worth of sample meals and even gave some recipes—all developed by the College of Home Economics at Cornell University and in some cases tested in the White House kitchen. In the chapter on recreation, she discussed the importance of a child obtaining a significant portion of his recreation outdoors, how exercise should be made into a pleasurable activity wherever possible, and the possibility of taking family walking tours in the American countryside similar to those that were commonly made by families in the Swiss Alps, the Scottish Highlands, and the English Lake District.

At her May 7, 1935 press conference, Eleanor invited Dr. Louise Stanley, head of the Bureau of Home Economics, a division of the Department of Agriculture, to talk about dried skim milk (Beasley 1983). Previously, when the butter was removed from raw milk, the remaining skim milk was discarded despite the nutrients it contained. The Department of Agriculture had developed a method for drying the skim milk, as well as inventing a strong waterproof sack for storing and transporting it that kept the milk fresh for 2 months. Stanley talked about the use of dried skim milk in home recipes and in commercial baking and ice cream production. Eleanor spoke about how the dried skim milk could be used to reduce the incidence of pellagra, a vitamin deficiency disease common during the Depression (Civitello 2011). The dried skim milk was distributed through federal and state relief programs. Eleanor's role in this case was not to lobby Congress, as in the case of farm surplus distribution, but instead to popularize an invention made by the USDA, encourage a change in industrial food production and distribution practices, and educate individual Americans how to avoid a common disease through inexpensive, vitamin-rich foods. In her typical fashion, Eleanor was much more concerned with scientific solutions to a common health problem than she was with the tastefulness of the foods made with this new ingredient.

5.2.3 Employment, Diet, and Exercise: CCC, TVA, REA, and Camp Tera

A number of organizations created through the first wave of New Deal legislation—the Civilian Conservation Corps (CCC), the Tennessee Valley Authority, and the Rural Electrification Administration—all had an impact on food policy. The CCC was created in 1933 to provide employment to men building roads, planting trees, and clearing trails. The impact on food policy was incidental. The work done by the CCC workers was physically demanding. The CCC offered the same daily rations as the U.S. Army, but the CCC administrators found they had to increase the rations because their workers were almost all undernourished. This experience provided useful data to the federal government in understanding an adult worker's daily food needs (Civitello 2011.)

The Tennessee Valley Authority, created in 1933, and the Rural Electrification Administration, created 2 years later, primarily were created to provide flood control, electric power generation, and rural power distribution. However, these two agencies helped to improve nutrition outside the cities by enhancing the food distribution network. They made available refrigerated storage lockers in which perishable foods could be stored prior to being sold to consumers, thus expanding the geographic range of distribution of fresh foods (Cummings 1940).

Eleanor wanted to build an employment program for unemployed women. In a press conference in 1934 she urged federal legislation to create forest work camps as a way to employ women, given that the CCC employed only men. There was not much support for this program among the Roosevelt Administration, but Eleanor persisted, including convening a White House Conference on Unemployed Women. Eventually she gained the support of Secretary of Labor Frances Perkins, and Camp Tera was established in New York State. Eventually, 90 women's camps, known as She-She-She camps, were run yearly, providing an outdoor experience for young adult women. The program, which served about 5,000 women per year, never came close to matching the CCC's scale of more than 200,000 men and was closed in 1937 (compared to the CCC camps, which persisted until 1942). Unlike the CCC, the women's camps did not provide wages or have the participants engage in physical work; instead the women received an outdoor experience, vocational counseling, and educational programs (Cook 1999; Kennedy 1999; Beasley 2010; "Great Depression," "Murray, Anna Pauline," and "Smith, Hilda Worthington" in Beasley et al. 2000; also read the press conference transcripts for July 6, 1933 and April 30, 1934 in Beasley 1983)

5.2.4 Social Security

The battle over the passage of a federal social security program was complex and had ramifications for American health care, partly but not only because it enabled the elderly population to have sufficient income to afford nutritious food and health care. National sickness insurance had been adopted in Germany in 1883 and in Britain in 1911. The Progressive movement had attempted to create national health insurance in the United States, but these efforts had foundered during the First World War. In debates over national health insurance during the 1930s, the discussions were often conflated with discussions of old age pensions and unemployment insurance. National health care was opposed in the 1930s by labor and business interests, and by the medical profession. Labor was opposed because they thought it might raise dues and reduce union membership. Doctors were opposed because it threatened professional autonomy and earning potential. Businesses were opposed because of the cost.

During the period of large job losses in the 1930s, old age pensions and unemployment insurance were regarded by most of the population as more important than health coverage. New Deal politicians did not want to jeopardize passage of Social Security legislation by insisting on health insurance as a component. The Social Security Act of 1935 authorized grants to states for programs targeting improved health of children and mothers, but it did not provide specific support for national health insurance or for sending federal funds to the states to establish such programs.

FDR was in favor of scientifically based health reforms, but the medical establishment represented a serious political obstacle. While the medical profession was opposed to national health insurance and generally any regulation limiting the practices or earnings of doctors, it was in favor of federal subsidies for medical care and research. For example, in its lobbying of Congress as the Social Security legislation was being drafted in 1934, the American Medical Association called for the continuation of private medical practice, control by doctors rather than lay people over professional medical standards and practices, freedom of doctors to choose their patients and methods of payment, and the ability of doctors to opt out of any health insurance schemes. The Roosevelt Administration studied various ways to support the medical field and American health more generally, but no bills for supporting national health insurance or subsidies to states to create health plans were passed during the Roosevelt years (Starr 1982; "National Health Insurance" in Beasley et al. 2000; Perrett 1985).

Eleanor lobbied in favor of passage of the Social Security Act of 1935. Typical of Eleanor's efforts in this area is a speech she gave before the D.C. Branch of the American Association for Social Security, the Council of Social Agencies, and the Monday Evening Club on February 8, 1934 (Roosevelt 1934). She called for a model social security program for the District of Columbia and gave a heartbreaking example of the health problems of a farm family through no fault of its own. In 1937 she urged Congress to extend the Social Security Act to protect farm workers, and on several occasions she urged for expansion of Social Security to include working women in traditional female occupations not then covered. She also argued for amendment of the Social Security legislation to include health benefits.

In addition to lobbying for more inclusive legislation, Eleanor worked to educate the American public about Social Security. She made multiple publicity visits to Social Security offices around the country ("Great Depression" in Beasley et al. 2000). She also worked to publicize the benefits available to mothers, widows, and children.

One unexpected development of the 1930s in response to the conservative medical establishment was an alternative health backlash, led by Bernarr Macfadden and his magazine *Physical Culture*. The magazine offered vegetarian menus, how-to advice on building muscles, articles touting the health value of fasting (seen as a process of detoxification), and claims for curing various diseases such as asthma through alternative means. As early as 1933, *Physical Culture* had circulation numbers in the hundreds of thousands of copies per month. The magazine included, for example, articles by movie star Clark Gable on the value of small breakfasts and Fay Wray (of King Kong fame) on the value of vegetable juice, and even a photograph of Eleanor Roosevelt eating pea soup with Macfadden at one of his whole-grain restaurants.

Eleanor held a business relationship with Macfadden. In 1932 she agreed to serve as the editor of a new monthly parenting magazine entitled *Babies—Just Babies* for Macfadden's empire of pulp magazines. She had apparently agreed to take on this role in order to provide a job for her daughter, Anna Roosevelt Dall, who was in the process of a divorce. She edited and wrote for the magazine for 6 months, but disagreements with Macfadden and ridicule from the Women's National Press Club caused her to resign her position. Not long thereafter, the magazine ceased publication (Beasley 2010). This alternative health movement died out with the coming of the Second World War and the demonstrable effect of the wonder drug penicillin (Adams 2009).

In order to coordinate and improve upon the patchwork of solutions created by the various acts passed in 1933 and the Social Security Act of 1935, FDR created the Interdepartmental Committee to Co-ordinate Health and Welfare Activities. It included 21 federal agencies, nine of them affiliated with the Department of Agriculture. The committee sponsored biochemical and physiological research to determine human nutritive requirements and checked these laboratory findings against actual consumption practices through economic and social studies. The findings served as the basis to calculate farm production requirements to meet the nation's nutritional needs. Government agencies used these findings when establishing goals for their own programs. The food processing companies vigilantly monitored this government activity in hopes of quashing any action that might negatively impact them.

When the Social Security Act of 1935 did not address health care, Eleanor and her friend Esther Lape brokered a meeting in 1937 between FDR and a group of physicians interested in federal support of medical education and research, as one step toward providing better care for the poorest third of the population. As a result, President Roosevelt convened a National Health Conference the next year to improve American health care. Senator Robert Wagner proposed legislation in 1939. However, opposition from the American Medical Association caused the Roosevelt Administration to drop the initiative ("Lape, Esther Everett" in Beasley et al. 2000). After the war Eleanor supported national health care, but she said little about it during the White House years, given the political sensitivity of national health insurance.

5.2.5 *Food Stamps, School Lunches, and Other Programs to Feed Poor Families*

Federal programs in the 1930s for food stamps and public school meals helped meet the nutritional needs of poor Americans. The food stamp program began in 1936. People on relief could buy orange food stamps at face value and use them for any food, and receive 50 cents worth of blue stamps free for every $1 of orange stamps purchased. Blue stamps could be used to buy only from a list of foods determined by the USDA, mostly surplus foods. The program ended in 1943, mainly because there was not much surplus food available during the war and many fewer people were unemployed. The highest number of people using stamps from that program at any one time was 4 million—at a time when the total U.S. population numbered about 135 million.

During the war, health and nutrition improved. On the home front, war workers often received hot meals and physical exams at company expense, and families had more income with which to buy nutritious foods. On the battle lines, although casualty and deaths for those at war were high, there was generally good food and improved health services. Despite various proposals by liberal members of Congress to reestablish a food stamps program during the Truman and Eisenhower administrations, it was not until 1961 that a new, trial food stamp program was created. It was made permanent by Congress in 1964 at the request of President Johnson (MacDonald 1977; New York Times 2010; US Department of Agriculture n.d., 2012; Cummings 1940; Center for the Study of the Presidency and Congress 2012; Perrett 1985).

It is unclear whether Eleanor Roosevelt had a role in urging Congress to create a food stamp program, although she had lobbied for reallocation of surplus food to poor families, as described earlier. In one of her radio broadcasts in 1942, she discussed with Secretary of Agriculture Claude Wickard the various federal programs that were important to poor families to keep up nutrition standards. These included the Food Stamp Program, the Penny Milk Program, and the School Lunch Program. She had held a press conference on November 8, 1939 in which she discussed the importance of another federal program to the nutrition of poor American families: the Farm Security Administration's program to resettle poor farm families to new farms where they could earn a living (Beasley 1983; also see the discussion later in this chapter of the resettlement program to Arthurdale, WV.)

The Roosevelt administration had a successful school food program that bolstered the role of the federal government in providing nutritious food to American children. In the nineteenth century, American schoolchildren were generally responsible for their own lunches. In the second half of the nineteenth century, in a few large cities, private charitable organizations helped with school lunches for the most needy children. For example, in 1853 The Children's Aid Society of New York started a program to feed students in vocational school. The importance of good nutrition to the mental and physical well-being of children was spelled out in Robert Hunter's 1904 book, *Poverty*, which identified his experiences in the Chicago and New York slums, and in John Spargo's 1906 book, *The Bitter Cry of the Children*.

As a result, the number of cities offering nutritious school lunches increased, and for the first time local governments or school boards assumed responsibility for these lunches. For example, in 1908 Philadelphia's school board took over the lunch program from charitable organizations. By 1937, 15 states had passed laws authorizing school boards to operate lunchrooms. In most cases the schools provided lunch at the cost of the food, but four states did reduce the cost for children of poor families (Cummings 1940; Levine 2010).

Unfortunately, these programs were too expensive for local governments or school boards to operate on their own. Federal help arrived in 1932, during the last year of Herbert Hoover's presidency, with support from the Reconstruction Finance Corporation (RFC), to pay the labor costs for preparing the school lunches. RFC was a federal agency that provided grants to states and local governments and loans to businesses, especially banks and railroads, so that they could continue to operate effectively. The Agricultural Adjustment Act, passed in 1933, enabled some of the surplus farm food bought by the federal government to be distributed to the schools for the lunch program. In 1935 the primary responsibility for the school lunch program shifted to another New Deal organization, the Works Project Administration. In some of the larger cities in the mid-1930s, WPA drivers would rush hot, freshly made meals prepared in centralized kitchens to schools throughout the city. By 1941 WPA workers were helping with school lunches in every state—not only in preparing and serving food, but also in building chairs and tables for the lunchrooms.

Through the federal food surplus program described earlier, by 1941 six million school children were being fed with this surplus food. However, once the United States entered the war, the amount of surplus food available to the schools dropped by 80 % because of the need to feed the military and the labor shortages on the farms. This did not cause a crisis, however, because adults were returning to work in record numbers as part of the war effort and thus there were fewer impoverished families.

During the final 3 years of the war, Congress made one-time allocations for school food purchases. Wanting to establish the program on a more secure, long-term basis, the National School Lunch Act was approved in 1946 under the Truman administration. The Act specified that federally subsidized lunches must meet minimum nutritional standards established by the Department of Agriculture. During the next 65 years, legislation tweaked the law in various ways but the program continues today and now serves over 30 million children.

5.2.6 Planned Communities

During the 1930s, when food was in short supply and nutritional requirements were hard to meet for many families, Eleanor became involved in various activities to improve the nutrition of working families. One initiative involved resettlement of workers to places where they could reside in decent homes and grow their own fresh fruits and vegetables. On behalf of the president, Eleanor made a visit in 1933 to the coal mining community of Scotts Run, West Virginia. The coal mining families were living in unsanitary conditions and general slum-like squalor, and they had trouble buying enough nutritious food. Eleanor convinced the president to back legislation, passed that same year, that provided funding for a Subsistence Homestead Division of the Department of the Interior. This division organized resettlement communities in rural locations, paid for with both federal and private philanthropic dollars. The first and most famous of these communities was located in Arthurdale, West Virginia. About 200 coal-mining families were resettled there—moving 30 miles from Scotts Run. Eleanor insisted the homes in Arthurdale be well constructed and have electricity, plumbing, and refrigerators. Because of the high cost of producing these houses and the fact that many other American families did not have these amenities in their own homes, the program was controversial. In one of her press conferences in 1934, Eleanor defended the program against charges that it was communistic and a waste of taxpayer dollars ("Arthurdale" and "Tugwell, Rexford Guy" in Beasley et al. 2000. Also see Eleanor's press conferences from April 11 and 23, 1934; January 30, 1936; May 29, 1939; May 13, 19, and 29, 1941; January 26, February 9, and May 20 1942—all in Beasley 1983).

Resettled away from their livelihood in the coal mines, the government intended for the people of Arthurdale to support themselves through farming and cooperative industries (not unlike the Val Kill cooperative manufacturing operation that Eleanor had earlier founded on the grounds of her private residence in Hyde Park, New York). West Virginia University Agricultural Extension Service ran an experimental farm at Arthurdale, where the ex-miners raised potatoes and other cash crops. An Appalachian craft center was opened in Arthurdale to preserve culture as well as generate income. However, an effort to use government funds to build a factory to construct post office equipment, as a further means to create jobs, was stymied by opponents in Congress who argued that this was antiprivate industry.

Eleanor took a strong personal interest in the everyday working of Arthurdale, but this caused some problems among the government bureaucrats who were administering the daily operation of the community. Members of the community would frequently go directly to Eleanor, bypassing the administrators in the Department of the Interior, and Eleanor repeatedly pressed government officials for quick action on various issues related to the community's operations, sometimes even taking issues directly to the president (Rexford Guy Tugwell in Beasley et al. 2000).

Fifty of these back-to-the-land resettlements were started by 1935, but the program was terminated several years later because the experiment was widely regarded as socialistic. Eleanor remained interested in the Arthurdale community in particular, which continued to operate with private funding, for example, attending the local school graduation every year through 1944.

5.2.7 Model Meals

Eleanor tried to improve nutrition during the Depression by serving as a role model and by suggesting model meals she had served at the White House. Her writings offered cooking tips, sample menus, and advice on balanced, inexpensive meals.

In the early years of the Depression, the problem was not a shortage of food but instead one of distribution and affordability. In the middle years of the Depression, drought and dust storms caused food shortages that persisted until the end of the 1930s.

Eleanor was determined to serve healthy, inexpensive meals to show the general populace how it could be done. Even before the presidential inauguration, she sought advice from the faculty at Cornell University, which operated a leading domestic science program that trained women in nutrition, chemistry, and sanitary engineering in preparation to be scientific homemakers. Eleanor was a firm believer in this scientifically based home economics movement. Food scientists at Cornell University developed three inexpensive foods—Milkorno, Milkoato, and Milkwheato—made from grain products, powdered milk, and salt—that were intended to help people stretch their budgets. These products were used as a base for sauces or to supplement meat dishes. In 1933, her first year in the White House, Eleanor directed the kitchen staff to serve Milkorno. For 1 week that year she also served economy lunches, costing only 7½ cents apiece, designed by the Cornell faculty for the Emergency Relief Administration. One of these lunches, for example, included hot stuffed eggs with tomato sauce, mashed potatoes, prune pudding, bread, and coffee. As was her wont, Eleanor was more concerned about nutrition and economy than about the attractiveness and tastiness of these foods. These meals were not seen as appropriate to the high position of the White House and thus were not served when important guests came to dinner. Some of the gestures toward food economizing in the White House were more modest, such as the use during the war of prunes to replace sugar at a time when sugar was scarce and expensive.

Eleanor was also interested in showcasing at the White House the finest local ingredients and regional dishes. While America had a rich food tradition, many American women at the time were cooking with canned soup, white bread, and American cheese. Eleanor brought in cookbook author Sheila Hibben to advise the White House kitchen on American culinary history, and Hibben introduced such items as stewed crabs, johnnycake, and chicory salad—many of which were taken from recipes that had been created for prior U.S. presidents. Hibben was interested in having food taste good, more than in making the feeding of the family an efficient and inexpensive process. She focused on local ingredients, skilled home cooking, and flavors of the past. She prepared 14 menus for the White House, each one with food products and dishes designed around the food of a particular region of the United States. For example, the New England menu included clam chowder, corned beef and cabbage, new potatoes, corn meal muffins, and Indian pudding. However, Eleanor's sensibilities and politics squared much better with the Cornell home economics school's values of economy, nutrition, and efficiency than it did with Hibben's focus on local harvesting of food ingredients, which were likely to be more expensive, not able to be gathered on a large enough scale to serve the nation, and often involving skilled cooking techniques. Eleanor shared the common American belief that it is perfectly acceptable for healthful, inexpensive food to be plain and tasteless (Shapiro 2010).

Hibben lost her influence in the White House and the Cornell recipes were too grim to serve as the basis for the White House menus on a regular basis. So Eleanor turned to her own housekeeper, Henrietta Nesbitt, to create a version of economical cookery. Nesbitt had no training as a professional cook. Eleanor had gone to the same church in Hyde Park as Nesbitt and previously had hired her to bake for the family in Hyde Park. Eleanor presented to Nesbitt on her first day of work at the White House a 28-item list of foods to be incorporated into meals for the president. It included whole wheat breads and unbleached flours, and suggested a local mill in Maryland as a source for stone-ground flour. In her White House memoir, Nesbitt referred to Eleanor's interest as one of "vitality-giving foods, plain and American" (Nesbitt 1948; also see Nesbitt 1951; Beasley 2010; Whitcomb and Mattielo 1998).

The meals Nesbitt planned were apparently wholesome but not appealing. Visitors in the know ate before attending a White House dinner. Ernest Hemingway said his White House meal in 1937 was the worst he had ever eaten. The same few dishes, many of them boiled dinners, were served repeatedly. The president detested Nesbitt's cooking, and Nesbitt frequently did not accommodate FDR's food requests. Nevertheless, FDR did not overrule Eleanor on what was served at the White House. Eleanor herself was not particularly interested in taking pleasure from food, and she recommended moderation in eating to others. She appreciated Nesbitt's frugality and trusted her, and was inured to the complaints about Nesbitt from FDR and others. When FDR's mother died in 1941, the President brought her cook (Mary Campbell) to the White House to prepare meals for him in a tiny kitchen on the top floor. But Nesbit drove Campbell out, arguing the foods she prepared were too rich for the President's health ("Nesbitt, Victoria Henrietta Kugler," "First Lady, Ceremonial Role," and "Alsop, Joseph W." in Beasley et al. 2000).

5.3 Formal and Informal Policy Approaches to Food and Nutrition During the Second World War

This section considers policy issues related to food during the Second World War. The formal policies of the Roosevelt Administration are intertwined with the informal efforts of Eleanor Roosevelt. Before turning to these specific programs, we consider the general policy environment for food during the first half of the 1940s.

The coming of the Second World War substantially changed the food policy environment in the United States. Spending for war production ended the recession. This happened even before the United States entered the war as a combatant in December 1941, for the nation had already become "an arsenal for democracy" to support the Allied forces. The new economic growth in the nation meant that refrigerators and automobiles were being bought in large numbers in 1940 as families finally had the funds to make these purchases. But inflation became a major threat to the return of a good middle-class life. When the 1941 government statistics indicated 12 % inflation in the cost of living, Eleanor bemoaned the increase in her magazine column. After the United States entered the war, conscription of farm workers led to shortages of food and increases in food commodity prices, further contributing to the inflation facing American families. President Roosevelt was worried about the impact of inflation and shortages on the American public; in his famous Four Freedoms speech in 1941, one of the freedoms he called for was freedom from want (Bentley 2002).

The government response was to increase its regulation of interactions between consumers and retailers, through both price controls and rationing, together with informal appeals to patriotism and sacrifices to ensure the American way of life. Business opposed both price controls and rationing throughout the war. War mobilization resulted in suspension or major reduction in production of various domestic goods such as refrigerators, irons, and personal automobiles because the metal was needed for tanks and bombers. The Office of Price Administration (OPA) was established in mid-1941 by the Roosevelt administration to set price ceilings on all goods other than farm products and to ration various goods such as tires, gasoline, and a number of foods eventually including sugar, coffee, and many kinds of meat. Price inflation was critically important in a time when wages were frozen and strikes were forbidden in industries contributing to the war effort. OPA was initially unsuccessful in keeping inflation in check, mainly because of inflation in agricultural products, so the Economic Stabilization Act was passed in 1942. It brought 90 % of food products under price control (Braverman 1996; Cohen 2003; Jacobs 2005; Perrett 1985; Ware 1942).

5.3.1 Kitchen Appliances as War Materiel

In 1941, before the United States had entered the war but after the U.S. Congress had passed the Lend Lease law that authorized the nation to provide war materiel to Britain, Russia, and other Allied nations, it was clear that there was a desperate shortage of aluminum for the construction of airplanes. In July the Federal Office of Production Management called for a 2-week nationwide scrap drive (Goodwin 1994). It was estimated that it would require 5,000 dishpans, 10,000 coffee percolators, 2,000 roasters, and 2,500 double boilers to have enough aluminum to build a single plane. The hope was that American housewives would provide enough scrap aluminum to build 2,000 planes. This was the first wartime call of sacrifice upon the American public, and it was not clear what the response would be. The response was overwhelming, and the drive far surpassed the target. Ironically, it turned out that scrap aluminum was not suitable for making planes, but it did serve other wartime purposes.

It is not clear what role, if any, Eleanor played in this particular drive. But the War Production Board organized many scrap drives during the war and Eleanor did promote a number of them. For example, in October 1942 she traveled to Seattle to rally support for a scrap metal drive being organized by three local newspapers (Royer 2011).

5.3.2 Nutritional Standards and Habits

Even before the United States entered the war, it was known that there would need to be some constraints on consumption on the home front in order to provide sufficient material aid to American and Allied soldiers. The Roosevelt administration hoped to determine a scientific basis that would establish how much food each solider needed so as to know how much food the nation should produce. While there was good scientific understanding about the need for the nutrients and vitamins provided by food, nobody knew exactly how much was needed. FDR convened a special nutrition conference in early 1940, and three nutritionists from the Food and Nutrition Board of the National Research Council were given a single day to write a complete set of nutritional standards. After a year of contentious argument, they arrived at recommended rather than required allowances; and where disagreement remained, they erred on the high side. Thus, these were daily requirements, not minimum daily requirements; many of the numbers were cautiously high, e.g., 3,000 cal per day for a 70-kg man. They were modified in 1944 and again several times after the war. Thus, no scientific basis for the food needs of the fighting troops was ever decisively determined (Cummings 1940; Bentley 2002).

Getting people on the home front to eat more healthily required changing food habits (Civitello 2011). In order to save meat for the fighting troops during the war, the government believed that the rationing program needed to be reinforced with a psychological campaign. While the Food and Nutrition Board handled nutritional standards, a Committee on Food Habits, led by anthropologist Margaret Mead and populated by social scientists, was charged by the National Research Council to study how to apply anthropological and psychological insights in order to change food habits and increase nutritional standards in the U.S. population. The group studied the social meaning of meals and mealtimes, especially during times of crisis, and it recommended small tables and family group settings in the lunchrooms. The Committee recommended that the government broadcast the basic message that to eat healthily was to display patriotism and help the war effort. The committee was disbanded after the war ended, and the government made no further formal efforts on the psychology of nutrition. However, the government did continue to offer nutritional guidance to the American people after the war through the Department of Agriculture and later the Department of Health and Human Services. Nutritional guidelines are updated by the Department of Agriculture every 5 years, based on the latest scientific research.

Eleanor spoke out on nutritional issues on various occasions during the war. At her April 29, 1942 press conference, for example, she pointed out the difficulties faced by temporary workers who had relocated to Washington, DC as part of the war effort in finding cafeterias to get their meals, especially if they had to work in the evenings. Three weeks later, at the May 19 press conference, she invited Dr. Louise Stanley from the Bureau of Home Economics together with Dr. Helen Mitchell, a professor of nutrition who was working during the war for the Social Security Administration, to talk about nutritional problems facing American families.

They discussed what the government was doing through the federal nutrition committee and urged local communities, especially in underprivileged neighborhoods, to become involved through nutrition institutes, women's clubs, and parent–teacher associations (Beasley 1983).

5.3.3 *Physical Education*

With increasing urbanization and industrialization in the nineteenth and early twentieth centuries, Americans generally got less exercise than they had when the majority of people worked on farms. For example, a U.S. Treasury Department Report on Exercise and Health in 1915 indicated that Americans were becoming too sedentary. Sports became popular in the late nineteenth century, and Teddy Roosevelt served as a role model for the nation as someone vital and physically fit. Nonetheless, one-third of recruits were found unfit for service in the First World War as a result of being underweight or having a nutritional disease such as rickets. Legislation after the war mandated better physical fitness programs in the public schools. But as the war became more distant, people lost interest in this issue; when money became tight during the 1930s, funding for school fitness programs was eliminated (Welch 1996; Rice et al. 1958; Dalleck and Kravitz 2002; Levine 2010).

As the United States began to prepare for the Second World War, the federal government took a renewed interest in physical education, with a particular interest in physical fitness and preparedness for military service. President Roosevelt appointed a former Olympic rower, John Kelly, to a newly created position as national director of physical training in 1940. When the United States entered the war and the Office of Civilian Defense was formed, Kelly took over responsibility for physical fitness within the OCD, while professional tennis player Alice Marble led the OCD's fitness program for girls and women (Berryman 1995; Park 1989; U.S. Surgeon General 1996). The Federal Security Agency established a large advisory body under the physician Charles Ward Crampton to advise the government on civilian physical fitness during the war (Crampton 1941; Park 1989). The American Medical Association also formed a committee in 1943 to study the role of exercise in physical fitness, under the direction of Chicago physiologist Arthur Steinhaus (Steinhaus et al. 1943). Wartime concerns about physical fitness focused primarily on muscle strength and growth, physical endurance, and disease resistance (Larson and Yocom 1951). Unfortunately, the Second World War was a reprise of the First World War, but with an even larger percentage of the recruits deemed unfit for service.

Eleanor had previously recognized the importance of physical activity as well as a good diet to promote healthiness. As mentioned earlier, she had already written on this theme in her book *It's Up to the Women* in 1933. Her efforts during the war to continue to promote physical fitness, however, landed Eleanor into some trouble.

Eleanor was keen to find roles for women to help with the war on the home front. Together with Florence Kerr, the director of the Work Projects Administration

Community Service Projects, Eleanor wrote a white paper that helped shape the creation of the Office of Civilian Defense. Eleanor originally stayed out of the operation of the Office, but when (September 1941) the director, New York City mayor Fiorello LaGuardia, did not do enough to involve volunteers in civil defense to satisfy Eleanor, she took a position as LaGuardia's unpaid assistant in order to push her agenda in the areas of nutrition, literacy, and physical fitness. (See Eleanor's implied criticism of LaGuardia in her press conference of September 3, 1941 in Beasley 1983.)

Eleanor used her position to secure for Mayris Chaney, a long-time friend and professional ballroom dancer, an appointment to a high-paying position as OCD's director of children's defense activities. Eleanor hoped that Chaney could institute a physical fitness program for children. However, there were complaints from political opponents of the Roosevelt administration, questioning Chaney's high salary and portraying as tawdry her background as a professional dancer. Eleanor resigned her position at the OCD not only because of Chaney's hiring and also her connection to the hiring of actor Melvyn Douglas, who was a Communist sympathizer paid a very high government salary, but also because of more general criticisms of conflict of interest between her appointment at OCD and her White House duties and access (Beasley 1987, 2010). Eleanor continued to defend Chaney's appointment in her Sunday evening radio broadcasts ("Office of Civilian Defense," Lash 1991; Beasley 1987; Goodwin 1994; "Congress," "Office of Civilian Defense," and "Democratic Party," "Morgenthau, Elinor" in Beasley et al. 2000; Eleanor's press conference of February 9, 1942 in Beasley 1983).

5.3.4 Rationing and Price Controls

In 1942, President Roosevelt proposed to Congress both rationing and price controls on various items. Rationing was not only about shortages but also about controlling prices and general inflation (Maddox 1992). There has been serious inflation during the First World War, and the Roosevelt Administration wanted to take a firm hand to restrain it this time around (Jacobs 1997). Later in the war, taxes were also increased as a means to keep inflation in check. Prices on controlled goods were set at the maximum price charged in March 1942. As mentioned earlier, the new federal agency OPA was created to both keep inflation in check and assure equitable distribution of materials in short supply. The OPA not only set price and rationing regulations, but it also attempted to set quality standards and grade products so that businesses did not try to get around the price controls by offering inferior products at the fixed prices set for standard quality products. This grading was similar to what the Department of Agriculture did with meat and eggs. However, the advertising industry and many manufacturers were opposed to grade labeling as restricting consumer choices, and by and large the OPA had little success in establishing quality standards and grading schemes.

As FDR explained to the American public in his second fireside chat over the radio, these measures of rationing and price control would provide "an equality of sacrifice" (Roosevelt 1942). The president needed to sell this program to the public because it was the most radical program until that time to regulate consumer consumption, and business was not supportive of his effort (Jacobs 1997). Each rationed item was given a price in points. Each person in the United States was given a book of stamps—worth 48 points per month and good for 6 months, which could be spent on any combination of rationed goods. Rationed food items included meat (but never turkey nor mutton), canned vegetables, sugar, coffee, and butter. Cigarettes, liquor, tires, and gasoline were also rationed. New tires were no longer sold unless one received a certificate of need for new tires. Blue points were given for processed foods; red points for meats, fats, and oils. Blue points were distributed monthly on an equal basis. Rationing coupons for shoes, sugar, and coffee were distributed on an interval rather than a monthly basis. Coffee allowance was provided only to adults. When a sale was made, the retailer would collect the points and use them to restock wares in the store.

Rationing was complicated and difficult to enforce. A total of 5,600 ration boards, staffed by volunteers, distributed the ration books each month at local schools to 130 million people. The limited availability of meat, butter, coffee, and sugar probably led to improved public health—as did the increase in walking caused by tire and gas rationing. Nobody starved, and most people suffered only mild inconvenience. However, rationing must have been frustrating to many American families. After almost a decade of Depression, when there was often not enough money to purchase basic necessities, many families were finally becoming better off financially, especially if they were working the enforced overtime hours in the defense plants, but the rationing system placed strict limits on what they could buy (Satterfield 1981).

The rationing system worked surprisingly well, considering the scale. Equal sharing plus special needs were the two guiding principles. People tried to work the system to argue special needs. Generally, people complied, but some coupons were bought and sold illegally on the black market. Numerous people tried to circumvent the gasoline and tire rules, which created the greatest hardship, despite fines as high as $10,000. Offenders of the rationing laws typically lost privileges rather than being fined or sent to jail (Satterfield 1981).

Price freezes were also difficult to enforce, and there were commonly efforts to game the system through bribes and the introduction of inferior products. OPA set prices on more than eight million items and affected the sales in more than three million business establishments. The ability to enforce these price controls outstripped the capacity of the 60,000 people who worked at the OPA. Neighbors would sometimes report people to the ration board, police, or the FBI. By the summer of 1944, much of the rationing had subsided, but OPA continued its operations into 1945 (Schenone 2003).

Women became an important political force during the war. Not only were they in the majority on the home front, they were also the major consumers for their families. The government called on consumers—most particularly female consumers—to watch and report merchants who were charging too high prices or consumers who were circumventing rationing laws. Eleanor helped to enlist women in this activity. In a publicity campaign, she was the first to sign the Home Front Pledge, eventually signed by more than 20 million American consumers, to not pay higher than legal prices for goods and not accept rationed goods without paying for them with ration coupons (Jacobs 1997, 2005; Ware 1942; Cohen 2003; Braverman 1996).

While there was some grumbling in American households about the rationing, coffee may have been the rationed food that was most acutely missed. Late in 1942, coffee was rationed to one cup a day for each person over age 15 because of the shortage of ships to carry the beans from South America. On average, the 83 million coffee drinkers in America had been consuming three cups a day. Eleanor tried to serve as a role model for American women, who were running the households and caring for the health, education, and welfare of their families. She carefully used ration coupons for her family, like any other family. She skillfully used the media to make sure that the public knew that she and the president received and used ration coupons in exactly the same way as any other American family, that she followed the coffee rationing rule in the White House, and that she even banned the traditional after-dinner demitasse for White House dinner guests. Eleanor did not care much about the effects of rationing on herself because she hardly noticed what she ate. She was concerned about what was served to the president and White House guests, but the most important thing was to assuage public dissatisfaction about rationing and reinforce the important role that women were playing on the home front. From time to time, Eleanor also tried to be a problem solver, just as she had with Milkorno during the Depression, by giving specific examples of how to make do in the face of shortages.

Eleanor actively supported rationing during the war. In 1942 she made 28 Sunday evening radio broadcasts sponsored by the Pan-American Coffee Bureau, in which she regularly urged women to accept food rationing. She wore black cotton stockings in public when silk came into short supply because it was needed for parachutes. On November 18, 1942, she talked about how America was in this fight together with the British, and how the British were making greater food sacrifices even than Americans (Beasley 1983; also see the August 25, 1941 press conference in which she discussed the food rationing and food shortages experienced by the Duke of Kent). Toward the end of the war, on April 2, 1945, she urged America to continue to produce food and not cut back in the face of possible surpluses in America because the rest of the world was in desperate need of food.

While most of her efforts to promote rationing were successful, one backfired. In a series of press conferences in 1942, Eleanor discussed sugar rationing. Sugar was used to make alcohol, an intermediate product in the manufacture of smokeless powder used by the military, but so far it had not been officially rationed. At her January 19 press conference, in an attempt to support the efforts of Secretary of Agriculture Claude Wickard, she encouraged Americans not to hoard refined white sugar and wherever possible to use alternatives such as brown sugar, raw sugar, molasses, and sorghum. One week later, at her January 26 press conference, she indicated that the White House would do its part to preserve sugar by cooking with products other than refined white sugar, and serving salads in place of deserts when sugar is not available. Sugar was also a topic for discussion at the February 9, April 29, May 14, and July 2 press conferences, at which Eleanor discussed government programs to ensure that restaurants and home canners had enough sugar for their needs, presented advice from the Department of Agriculture on how to can fruits using less sugar, and noted how she would tell guests visiting her home at Hyde Park to bring their own sugar with them. Although Eleanor's intention was to serve as a role model who promoted self-rationing of sugar and as a provider of coping strategies, the American public believed—at least partly from her remarks—that a shortage of sugar was imminent. There was a run on sugar in the stores, which led to an acute shortage followed by official rationing. With careful management of the sugar available, formal rationing might otherwise have been avoided (Beasley 1983).

5.3.5 Victory Gardens

Victory Gardens had existed during the First World War, including a million gardens grown by children at school at the urging of President Woodrow Wilson. When the Second World War began, people began to grow their own fruits and vegetables. More interested in industrial-strength solutions than the haphazard efforts of millions of individuals, and perhaps because of an effort to protect agribusiness interests, the Department of Agriculture was not enthusiastic about these Victory Gardens at the beginning of the Second World War. A different federal agency, the Farm Security Administration, began what proved to be a successful program in rural areas, known as An Acre for a Soldier program. Crops raised on or profits from that acre were donated to supply canteens for servicemen. Farmers who did not have land to spare donated a pig or sheep to the program (Perrett 1985). Moreover, American families began to plant their own gardens without encouragement from the USDA. By April 1942 more than six million individuals had planted Victory Gardens. Since they could not stop these efforts, the Department of Agriculture tried to organize them. Claude Wickard, the Secretary of Agriculture, called later that year for 18 million Victory Gardens. The growing conditions were good that year, and 1942 saw bountiful crops, which helped to offset the anticipated food shortage created by the many farm workers who were off to war. Volunteers from the YWCA, High School Victory Corps, and American Women's Voluntary Service helped to bring in the harvest.

For the remainder of the war, Victory Gardens were a major part of the home front war effort. Towns dug up parks for use by amateur gardeners—as did prisons, high school and college campuses, and convents. At its main store in Manhattan, Macy's began selling ducks, chickens, and rabbits as well as coops to house them. The eight million tons of food from Victory Gardens in 1943 went a long way toward reducing anticipated shortages. The government worked with the radio networks and the advertising agencies to promote home Victory Gardens. Radio drama series brought children into the war effort. For example, the Jack Armstrong Write-A-Fighter Corps (associated with the radio show *Jack Armstrong, the All American Boy*) enlisted more than a million boys to write letters to servicemen overseas, collect scrap metal, sell war bonds, and work in Victory Gardens. One canning company, working in cooperation with the federal government, helped the cause of Victory Gardens with an ad headlined: "Wanted: 1,000,000 Competitors." More than 50 million Victory Gardens were planted during the war; 40 % of all fresh vegetables consumed by civilians during 1944 came from Victory Gardens.

There were several calls for a victory garden at the White House. Eleanor had encouraged the public to grow vegetables to help avoid a national food shortage, especially after the USDA became supportive of Victory Gardens; she was in favor of growing vegetables on the White House lawns although she did express uncertainty about another proposal made at one of her press conferences to let sheep roam the White House lawns—something that the Wilsons had done during the First World War. In the early years of the war, the Department of Agriculture discouraged city dwellers from planting victory gardens because the soil was not good enough to make efficient use of seeds and fertilizer, which were in short supply. In 1942 the Department of Agriculture determined that the ground under the White House lawn was unsuitable for a victory garden, and none was attempted that year. However, it was eventually determined that the flower beds were a more suitable place for a vegetable garden, and a small Victory Garden was planted in 1943 (Eleanor Roosevelt Press conferences, April 6 and 29, 1942 in Beasley 1983; Roosevelt Library 2011).

5.3.6 *Eleanor Clubs*

Eleanor Roosevelt was a strong supporter of civil rights. She spoke out on behalf of the employment conditions of Black domestic workers. Not surprisingly, she was a target of fear and hate in the South. Apparently it was white supremacists who spread rumors heard in the early 1940s that she had supported the formation of "Eleanor Clubs" to unionize black cooks and maids and encourage them to leave kitchens of white families unless they received better pay and more equal treatment. At Eleanor's insistence, these allegations were investigated by the Federal Bureau of Investigation in 1942 and found to have no substance. It was, no doubt, difficult to retain Black help in the home during the war, when better paying war factory work was available (Beasley 1987, 2010; "Eleanor Clubs" in Beasley et al. 2000).

5.4 Conclusions

The Roosevelt years in the White House were active ones for food policy. Families could not afford the food that was being grown during the Depression, creating dire circumstances for both poor families and farmers. Food was in short supply during the war years because of the need to feed the troops, the lack of transportation for importing food, and the farm labor shortage due to conscription. The Roosevelt administration offered many formal food programs: the Agricultural Adjustment Act and surplus food reallocation; food and nutritional education through the USDA; activities of the CCC, Tennessee Valley Authority, and Rural Electrification Authority in the areas of diet and exercise; a food stamp program; planned communities such as Arthurdale, where poor people could grow their own food; nutritional standard setting through efforts of the National Research Council; and a federally promoted national fitness campaign.

Eleanor Roosevelt played an active informal role in using the bully pulpit of the White House to promote food, nutrition, and physical fitness. She was an adept and untiring user of the media to spread her message, not only through traditional press conferences and public speeches, but also through radio broadcasts and newspaper and magazine articles and columns. She carefully cultivated a public persona that would appeal to both stay-at-home and working women, and she crafted her messages to appeal to women as the keepers of home and society.

Eleanor's informal food policy was carried out in at least five ways. She worked to influence legislation, such as the federal distribution of surplus farm goods in the early years of the Depression. She encouraged federal agencies to expand their food and health programs, such as her work on behalf of the model community at Arthurdale, Camp Tera for poor women, and her various efforts at the Office of Civilian Defense. She worked to change individual habits, such as her writings directed at American mothers providing advice on diet and physical exercise for their families. She provided coping strategies during times of stress, including sample budget-stretching meals during the Depression. She also offered reassurance to individual Americans, such as her wartime messages about rationing. Her efforts at informal food, health, and nutrition policy were unparalleled by any First Lady until the arrival of Michelle Obama in the White House. It is too soon to know which of them will have had a greater impact on food policy.

References

Adams M (2009) Smart goop: Depression-era health regimens return. New York. 42(10):10–11

Beasley M (ed) (1983) The White House press conferences of Eleanor Roosevelt. Garland, New York

Beasley MH (1987) Eleanor Roosevelt and the media. University of Illinois Press, Urbana

Beasley MH (2010) Eleanor Roosevelt: Transformative first lady. Lawrence: University of Kansas Press.

Beasley MH, Schulman HC, Beasley HR (eds) (2000) The Eleanor Roosevelt encyclopedia. Greenwood, Westport

Bentley A (2002) Islands of serenity: Gender, race, and ordered meals during World War II. In: Counihan CM (ed) Food in the USA. Routledge, New York, pp 171–192

Berryman JW (1995) Out of many one: a history of the American College of Sports Medicine. Human Kinetics, Champaign, Illinois

Black AM (1996) Casting her own shadow: Eleanor Roosevelt and the shaping of postwar liberalism. Columbia University Press, New York

Braverman J (1996) To hasten the homecoming: How Americans fought World War II through the media. Madison Books, Lanham

Center for the Study of the Presidency and Congress (2012) The history of SNAP. SNAPtoHealth.org. http://www.snaptohealth.org/snap/the-history-of-snap. Accessed 16 Nov 16 2012
Chase S, Schlink, F (1927) Your money's worth: a study in the waste of the consumer's dollar. Macmillan, New York
Civitello L (2011) Cuisine and culture, 3rd edn. Wiley, Hoboken
Cohen L (2003) A consumer's republic: The politics of mass consumption in postwar America. Knopf, New York
Cook BW (1999) Eleanor Roosevelt: Volume two, 1933–1938. Viking Press, New York
Crampton CW (1941) Start today: Your guide to physical fitness. A.S. Barnes, New York
Cummings RO (1940) The American and his food. University of Chicago Press. Chicago
Dalleck LC, Kravitz L (2002) The history of fitness. IDEA Health and Fitness Source. http://www.ideafit.com/fitness-library/the-history-of-fitness. Accessed 19 Dec 2013
Denker J (2003) The world on a plate. University of Nebraska Press, Lincoln
Dennis PM (1995) Between Watson and Spock: Eleanor Roosevelt's advice on child-rearing from 1928 to 1962. J Am Cult 18(1):41–50
Elias MJ (2009) Food in the United States, 1890–1945. Greenwood, Santa Barbara
Goodwin DK (1994) No ordinary time: Franklin and Eleanor Roosevelt; The home front in World War II. Simon & Schuster, New York
Hickok L(1962) Eleanor Roosevelt: Reluctant first lady. Dodd Mead, New York
Hunter R (1904) Poverty. Macmillan, New York
Jackson J (1968) The Ergot controversy: Prologue to the 1938 Food, Drug, and Cosmetic Act. J Hist Med Allied Sci XXIII:248–257
Jacobs M (1997) 'How about some meat?': The office of price administration, consumption politics, and state building from the bottom up, 1941–1946. J Am Hist 84(3):910–941
Jacobs M (2005) Pocketbook politics: Economic citizenship in twentieth-century America. Princeton University Press, Princeton
Kallett A, Schlink F (1933) 100,000,000 Guinea pigs: Dangers in everyday foods, drugs, and cosmetics. Vanguard, New York
Kennedy D (1999)Freedom from fear: The American people in depression and war, 1929–1945. Oxford University Press, New York
Larson LA, Yocom RD (1951) Measurement and evaluation in physical, health, and recreation education. C.V. Mosby, St. Louis
Lash JP (1991) Eleanor and Franklin. Norton, New York
Levenstein H (2003) Revolution at the table. University of California Press, Berkeley
Levine S (2010) School lunch politics: The surprising history of America's favorite welfare program. Princeton University Press, Princeton
MacDonald M (1977) Food stamps: An analytical history. Soc Sci Rev 51(4):642–658
Maddox RJ (1992) The United States and World War II. Westview, Boulder
Mayer RN (1989) The consumer movement. Twayne, Boston
McGovern CF (2006) Sold American: Consumption and citizenship, 1890–1945. University of North Carolina Press, Chapel Hill
Nesbitt H (1948) White Housed Diary. Doubleday, New York
Nesbitt H (1951) The Presidential cookbook. Doubleday, New York
Nestle M (2007) Food politics. University of California Press, Berkeley
New York Times (2010) A history of food stamps use and policy. http://www.nytimes.com/interactive/2010/02/11/us/FOODSTAMPS.html. 11 Feb 2010, Accessed 19 Dec 2013
Ocepek MG, Aspray W, Royer G (Forthcoming) The incidental impact of first lady Jacqueline Kennedy on the revival of french food in America. Food Stud: Interdiscip J
O'Farrell B (2010). She was one of us: Eleanor Roosevelt and the American worker. Cornell University Press, Ithaca
Park RJ (1989) Measurement of physical fitness: A historical perspective. Office of Disease Prevention and Health Promotion Monograph Series. U.S. Department of Health and Human Services, Public Health Service, Washington DC

Parran TD (1940) Results obtained on the medical appraisal of the state of nutrition. In: Proceedings of the 18th Annual Conference of the Milbank Memorial Fund, pp 39–43

Perrett G (1985) Days of sadness, years of triumph: The American people, 1939–1945. University of Wisconsin Press, Madison

Rice EA, Hutchinson, JL, Lee M (1958) A brief history of physical education. Ronald Press, New York

Roosevelt E (1933) It's up to the women. Stokes, New York

Roosevelt E (1934) Old age pensions. The Eleanor Roosevelt Papers Project. http://www.gwu.edu/~erpapers/documents/articles/oldagepensions.cfm. 8 Feb 1932, Accessed 14 Dec 2013

Roosevelt E (1958) Hostess at Hyde Park. Saturday Evening Post. 15 Feb 1958, pp 32–33, 106–108

Roosevelt FD (1942) A call for sacrifice. Modern History Sourcebook. http://www.fordham.edu/halsall/mod/1942roosevelt-sacrifice.html. 28 April 1942, Accessed 25 Sept 2013

Roosevelt Library (2011) Found in the archives: Victory gardens and shared sacrifice. http://fdrlibrary.wordpress.com/tag/victory-gardens/. 18 Aug 2011, Accessed 20 Oct 2013

Royer B (2011) Garbage in. The Cascadia Courier. http://www.thecascadiacourier.com/2011/07/garbage-in.html. 25 July 2011, Accessed 28 Aug 2013

Satterfield A (1981) The home front: An oral history of the war years in America, 1941–1945. Playboy Press, Chicago

Schenone L (2003) A thousand years over a hot stove. Norton, New York

Shapiro L (2010) The first kitchen. New Yorker. 22 Nov 2010, pp 74–80

Spargo J (1906) The bitter cry of the children. Macmillan, New York

Starr P (1982) The social transformation of American medicine. Basic Books, New York

Steinhaus AH, Bauer WW, Hellerbrandt FA, Kranz L, Miller A. 1943. The role of exercise in physical fitness. J Health Phys Educ Recreat 14:299–300, 345

Tugwell R (1963) Remarks. Roosevelt Day Dinner Journal Americans for Democratic Action. In: the Eleanor Roosevelt Papers Project http://www.gwu.edu/~erpapers/teachinger/q-and-a/q20.cfm. 31 Jan 1963, Accessed 30 Aug 2013

U.S. Department of Agriculture (2012) Supplemental Nutrition Assistance Program: A short history of SNAP. Food and Nutrition Service. http://www.fns.usda.gov/snap/rules/Legislation/about.htm. Accessed 16 Nov 2012

U.S. Department of Agriculture. (nd) From food stamps to the Supplemental Nutrition Assistance Program: Legislative timeline. http://www.fns.usda.gov/snap/rules/Legislation/timeline.pdf. Accessed 16 Nov 2012

U.S. Surgeon General (1996) Physical activity and health: A report of the Surgeon General. http://www.cdc.gov/nccdphp/sgr/index.htm. Accessed 19 Dec 2013

Ware C (1942) The consumer goes to war: A guide to victory on the home front. Funk & Wagnalls, New York

Welch PD (1996) History of American physical education and sport, 2nd edn. Thomas, Springfield

Whitcomb C, Mattielo M (1998) Your invitation to Mrs. Roosevelt's table. Vic 12(6):74

Winfield BH (1990) The legacy of Eleanor Roosevelt. Pres Stud Q 20(4):699–706

Chapter 6
Food Policy Since 2009: The Obama Administration's Policies and Michelle Obama's Bully Pulpit

Some kids have never seen what a real tomato looks like off the vine. They don't know where a cucumber comes from. And that really affects the way they view food. So a garden helps them really get their hands dirty, literally, and understand the whole process of where their food comes from. And I wanted them to see just how challenging and rewarding it is to grow your own food, so that they would better understand what our farmers are doing every single day across this country and have an appreciation for … that American tradition of growing our own food and feeding ourselves.

(Michelle Obama at the U.S. Department of Agriculture, 5/3/13, Holecko 2013)

As a mom, I know it is my responsibility—and no one else's—to raise my kids. But what does it mean when so many parents are finding that their best efforts are undermined by an avalanche of advertisements aimed at their kids? And what are these ads teaching kids about food and nutrition? That it's good to have salty, sugary food and snacks every day—breakfast, lunch, and dinner?

(Michelle Obama at the Grocery Manufacturers Association, 3/16/2010, Holecko 2013)

I forgot what election she [Michelle Obama] won again? Oh wait, she never ran for any office that gives her the authority to do this. So while the First Socialist of the United States forces her new foods laws down our throats, she is out there piging [sic] out too. Have you seen the amount of unhealthy food her [sic] and her husband eat? Do as I say, not as I do apparently.

(D'Andrea 2012)

W. Aspray et al., *Formal and Informal Approaches to Food Policy*, SpringerBriefs in Food, Health, and Nutrition, DOI 10.1007/978-3-319-04966-3_6,

Chapter 5 discussed how Eleanor Roosevelt used the bully pulpit of the White House to advance goals of better food, health, and nutrition for the American public. The years of the Roosevelt presidency were times of want, coeval with most of the Great Depression and the Second World War. This chapter tells a similar story, of Michelle Obama's use of the bully pulpit of the White House, also for purposes of improving food, health, and nutrition of the American people. The striking difference in these two cases is that the Obama presidency is occurring in a time of plenty, when there is a major national problem with obesity. Similar to Chap. 5, this chapter interleaves the informal actions of the First Lady with the formal actions of her husband's administration. We have the advantage of distance in time from the Roosevelt story, which helps us to gain historical perspective—an advantage we do not have in the Obama study. Thus, this chapter is necessarily more tentative and less complete than the Roosevelt chapter.

Concerns about issues of public health and their economic and national defense implications have helped to shape Michelle Obama's activities in the White House. Her background working in hospitals may have influenced how she has chosen to use her position as First Lady. She was born into a working-class family in Chicago and excelled as a student, attending Princeton University and Harvard Law School. While working for the Chicago Law firm Sidley Austin she met her husband, who was one of the summer interns she mentored. Later she worked as a senior manager in community relations for the University of Chicago Hospitals. Her family continued to live in Chicago while Barack served as U.S. Senator. When the family moved to the White House, Michelle decided that while she would work on policies and practices supporting working families and military spouses, her first priority would be to improve nutrition and decrease obesity among America's children.

The Obamas have two daughters, and Michelle often expresses her goals related to nutrition and obesity in the terms of a mother who wants to provide proper care for her family. Many times she has told the story of a pediatrician who reported to her that one of her daughters was overweight and how, at the pediatrician's recommendation, she began to feed both of her daughters a less fattening diet. The family had fallen into a common routine that involved eating out, ordering pizza for delivery, and making sandwiches because of the family's busy schedules. In order to change the family's diet, Michelle began to serve more fruit and vegetables, prepare meals several times a week instead of eating out all the time, and eliminate some processed and sugary foods. These small changes, she reported, had immediate results. This style of eating also benefited her husband, she noted, who had been overweight as a child and as an adult avoids fatty foods. While what was happening within her own family was likely a powerful force in helping decide what issues to emphasize while in the White House, it is not lost on the First Lady how powerful a rhetorical strategy it is to use this message about a mother trying to care well for her family as she reaches out to policy makers and the American public.

Barack Obama's presidential platform meshes nicely with Michelle's interest in nutrition and obesity. His highest policy priorities when he entered the White House were health, energy, and education. He has likened himself to Teddy Roosevelt, and he and Teddy Roosevelt are similar in distrusting bureaucracy, wanting to make government more efficient, and using public scrutiny to rein in government practice (Lynn 2009; and for a criticism of the Obama-Teddy Roosevelt comparison, see Fournier 2011). During his presidential campaign, Obama produced a policy paper on food and agriculture reform entitled "Real Leadership for Rural America." It calls for people to buy fresh and local products, expand organic farming through federal subsidies, encourage careers in farming and ranching, provide nutrition education in schools, and reduce the amount of energy used in industrial food production.

Michelle Obama's campaign on nutrition and obesity has been carried out mainly through two efforts: the creation of a garden at the White House and the Let's Move campaign, which is based upon four principles: better nutrition information, increased physical activity, easier access to healthy foods, and personal responsibility (Holecko n.d.). She has become a powerful figure in America's fight against obesity. She shows up each year near the top of the list of America's 50 Most Powerful People in Food, and she was the winner in 2011 of the James Beard Leadership Award. Her method is largely to use the bully pulpit, but she has also worked closely with the formal policy efforts of the Department of Agriculture and the Institute of Medicine (Gavin 2012). One dimension of the obesity issue that she has avoided is direct confrontation with the food industry over television and Internet advertising of harmful foods. (See Chap. 3 for this history.)

6.1 The White House Garden

The first major symbolic act in Michelle Obama's campaign against obesity was the planting of a garden at the White House. In March 2009, as soon as the ground began to thaw and less than 2 months after the presidential inauguration, the South Lawn of the White House was tilled for a fruit and vegetable garden. While most of the work to till the soil, and later to plant, fertilize, maintain, and harvest the vegetable beds was done by the White House ground crew, some of the labor was provided by fifth-grade students from a local school and by the First Family and volunteers from the White House staff. The intention was to use the garden to produce healthy ingredients for the meals of the First Family and state dinners. The excess food was to be donated to Miriam's Kitchen, a District of Columbia food kitchen that feeds the homeless and uses only fresh ingredients (Swarns 2009). The garden is also intended to teach children, and through them their families, about the value of healthy, locally grown fruits and vegetables. In its first year, the garden produced over 1,000 lb of vegetables, fruits, and herbs as well as 134 lb of honey (Black 2009; Burros 2009b, 2012).

The Obama garden follows a longstanding, if only occasional tradition of using the White House grounds for agricultural and educational purposes. John Adams, the first president to live in the White House, planted a garden soon after taking residence in 1800. Woodrow Wilson grazed sheep on the White House lawns in 1918 to keep them trimmed and fertilized, so as to conserve resources for the war effort.

President Wilson's Administration convinced a million children to plant gardens at home to lessen food shortages during the First World War. The Roosevelts had a small Victory Garden on the White House lawn during the later years of the Second World War. Jimmy Carter, who had a background as a peanut farmer, spoke of the virtues of gardening during his campaign but never planted a garden at the White House despite calls to do so. The Clintons approved a small garden in pots on the roof, but they declined requests for lawn gardening, arguing that it would not fit with the formal nature of the White House lawns.

A number of individuals lobbied the Obamas to create a White House Garden. One was Roger Doiron, the founder and director of Kitchen Gardeners International. He circulated a petition entitled "Eat the View," which in January 2009 beat out 4,000 other entries in the "On Day One" contest from the United Nations Foundation. Michael Pollan, a professor of journalism at Berkeley, food activist, and author of *The Omnivore's Dilemma*, published an open letter to the president, entitled "Farmer-in-chief" in which he suggested a move back to sun-based food and away from industrially grown foods. Pollan argued that agribusiness is the second highest user of fossil fuels after automobiles, 37 % of greenhouse emissions come from industrial food producers, and industrial food production is highly inefficient and wasteful of energy.

A third call came from Daniel Bowman Simon, who drove across the country in a school bus with a renewable garden on its roof to promote a White House garden. Simon and a colleague called for use of heirloom seeds and for compost to be made from the waste collected from the White House, Congress, and Supreme Court kitchens. Simon came to the attention of the Obamas through Sam Kass, who had been their personal chef in Chicago—the person who had the charge of weaning the Obama daughters from fast food and keeping their weight under control. Kass joined the Obama White House kitchen staff and became a senior policy advisor for the healthy food initiative in the Obama administration. He had met Simon through Rethinking Soup, an organization that Kass had helped to found, which brings famers and activists together to discuss food issues.

Perhaps the most vocal encouragement came from Alice Waters, the owner of the famous Chez Panisse restaurant in Berkeley, California, which popularized the use of organic, locally grown foods. In 1995 Waters had begun working with the Martin Luther King, Jr. middle school in Berkeley. Their partnership converted the school's 1 acre of blacktop into a garden and kitchen classroom, as explained on her Edible Schoolyard website (edibleschoolyard.org). The plot was used not only to grow a bountiful harvest of vegetables, fruits, herbs, and flowers, but also to teach almost 1,000 middle-school students each year about raising and cooking healthy foods. In 2005, through her Chez Panisse Foundation, Waters opened a similar program in New Orleans and then later in Greensboro, North Carolina and Brooklyn, New York (Salter 2010).

Waters had been lobbying for a White House garden since the first Clinton administration. Punning on the term first used in the nineteenth century for unofficial advisors to President Andrew Jackson, Waters recommended the establishment of a "Kitchen Cabinet"—an advisory committee to help hire a White House executive chef who would pay attention to environmental, health, and conservation issues. *Gourmet* magazine's Ruth Reichl and New York restaurant owner Danny Meyer volunteered to join Waters as members of the Kitchen Cabinet (Bailey 2009). They suggested some names to hire as White House chef, including Art Smith (Oprah Winfrey's personal chef) and Rick Bayless (chef at Chicago's Topolobampo, a favorite restaurant of the Obamas). However, Michelle Obama chose instead to retain Cristeta Comerford, who had served George W. Bush and was the first female to serve as White House executive chef. The Kitchen Cabinet expressed concern over this choice, given the notoriously poor diet of President Bush, but later they learned that Comerford had been serving organic meals to the Bush family all along.

One of the few groups not in favor of the garden was the Mid-America Crop-life Association. It argued that the White House's effort to get people to produce their own food is misguided. Instead, the Association emphasized the virtues of industrialized food production and asked: "if Americans were still required to support their family's basic food and fiber needs, would the U.S. have been leaders in the advancement of science, communication, education, medicine, transportation and the arts?" (McCarvel 2009 as quoted in Batra-Wells 2010).

The garden of 1,100 ft^2 is visible from the street, through the White House fence. It is a stop on the living history public tour of the White House and thus is visited by more than a million people each year. The White House chefs selected 55 varieties of vegetables to grow, as well as selections for the berry and herb patches. The gardeners use local nutrients for fertilizer such as White House compost, crab meal from the Chesapeake Bay, lime, and green sand. Ladybugs and praying mantises are used to provide organic bug control. The two beehives in the garden pollinate the plants as well as producing honey.

The selection of plants and gardening practices has undergone close cultural scrutiny. The vegetables include a number of heirloom varieties, taken from Thomas Jefferson's home at Monticello, such as Tennis Ball lettuce, Marseilles figs, and Case Knife pole beans. Cultural critics, such as Puja Batra-Wells, argue that this selection harkens back to an agrarian ideal represented by Jefferson that contrasts with industrialized eating. It emphasizes Jeffersonian ideals such as self-reliance, making use of one's own resources, and taking responsibility for one's life. It rallies against consumption of mass-produced processed foods shipped long distance, ecological problems created through industrial production methods, and overconsumption through commercial advertising and product portioning. Batra-Wells notes that Michelle Obama talks about the gardening choices not in terms of being organic and sustainable, but instead as family choices that help her family to eat fresh and nutritious foods that help stave off obesity. This rhetoric Batra-Wells interprets as a way to avoid criticism from the agricultural and industrial food industries. The use of the White House garden as a pedagogic device is not new with Michelle Obama. It is a technique that harkens back to the period from 1890 to 1920, when gardens were used to teach natural science, agriculture, and esthetics of nature (Batra-Wells 2010; also see Cameron 2012).

The cultural discussions of the White House garden also concern issues of class. For example, Batra-Wells argues that the foods selected for planting in the White

House garden are specialty, upper-middle and upper class foods, not foods that would be eaten by the general public. These include, for example, arugula, sorrel, Thai basil, Marseille figs, tarragon, and chervil. This selection, it turns out, conforms well with the Obamas' eating preferences such as artisanal chocolates and gourmet Mexican restaurants. Much has been made of an incident in Adel, Iowa, when Barack Obama was on the campaign trail in his first presidential election. He told farmers at the Rural Issues Forum that he understood their plight, of supermarket prices rising with little of that increase returned to the farmers who grow the crops: "Anybody gone into Whole Foods lately and see what they charge for arugula?" (as quoted in Batra-Wells 2010; also see Kantor 2007 about other food problems on the campaign trail). The irony is that the specialty grocer Whole Foods does not operate any stores in Iowa, and no farm in Iowa grows the specialty salad green arugula.

In order to popularize the garden and its pedagogic value, Michelle Obama published a cookbook book, *American Grown: How the White House Kitchen Garden Inspires Families, Schools, and Communities* (Obama 2012). It presents recipes with fresh foods, describes how to garden in various settings, and tells the story of the White House garden. (On the rise of community gardens in America, see Todd 2009.)

6.2 Let's Move!

In February 2010, less than a year after work began on the White House garden, Michelle Obama announced the Let's Move initiative (sometimes called the Healthy Kids Initiative) with the goal of solving within a generation the obesity epidemic among America's children. She linked her obesity campaign to public health reform, arguing that obesity-related disease costs the economy $147 billion each year. The program has four components, and these four components have defined her activities in the fight against obesity in the White House:

- *Healthy Choices*: better nutrition labeling, a revamped food pyramid, and regular monitoring of the body mass index (BMI) of children.
- *Healthy Schools*: reauthorization of the Child Nutrition Act for improved school lunches and expansion of the Healthier US Schools Challenge, a 2004 initiative to recognize schools that are reaching certain goals concerning promotion of nutrition and physical education.
- *Physical Activity*: revamping the President's Physical Fitness Challenge program and increasing participation, and enlisting professional athletes to encourage kids to exercise 60 min per day.
- *Access to Affordable Healthy Food*: eliminating urban and rural food deserts.

Let's Move has received bipartisan political support, as well as support from the medical profession and the food industry. To punctuate the launch, President Obama signed an executive order to create a national task force on childhood obesity, drawing on the resources of the federal departments of Interior, Health and Human Services, Agriculture, and Education (Givhan 2010). He also pledged as much as $1 billion per year for 10 years in federal funds. The Food and Drug Administration (FDA) was assigned to work with food and beverage manufacturers, who have agreed to improve package labeling. The goal is to make ingredients more prominent and easier to understand on the packaging and to make it easier to figure out calorie and other nutrition facts for a normal size serving. Nutrition in public schools is being addressed at the federal level through reforms included in the reauthorization bill for the Child Nutrition Act, which covers school lunches. Companies that supply food materials to schools, including Aramark, Sodexo, and Chartwell School Dining Services, have agreed to cut salt and fat and add more whole grains and more fresh fruits. (See the account by the former head of the FDA: Kessler 2009. Also see Chap. 4.)

In 2009 Michelle Obama hired Sam Kass for a policy position, as coordinator of the food initiative, with the goal of getting the entire country to eat better. Kass trained professionally as a chef in Austria and cooked at Chicago's Avec restaurant before becoming the Obamas' personal chef in Chicago. In 2011 Michelle hired Judith Palfrey, a pediatrician and former head of the American Academy of Pediatrics, to be executive director of Let's Move. Michelle also announced the creation of a new foundation, called the Partnership for a Healthier America, which is being supported by three large health care organizations: the Robert Wood Johnson Foundation, Kaiser Permanente, and the W.K. Kellogg Foundation.

As First Lady, Michelle Obama cannot expect large television audiences if she calls press conferences. Even sitting presidents have had difficulty gaining the public's attention since the advent of cable television. So she has had to use other means to get the word out about her obesity campaign. She gives lectures to groups all over the country. She employs the print media. For example, she appeared on the front cover of *Vogue* magazine, discussed her and the president's morning exercise workouts in *People* magazine, and told *Parents* magazine in a published interview how she and her husband had eliminated juice boxes and cut back on processed food for their family (Westfall 2009; Alina 2011).

She also uses television. For example, she appeared on *The Today Show* and *Live With Regis and Kathy* on consecutive days in 2011 (Stolberg 2011). To get the attention of the late night crowd, she appeared on *The Tonight Show*. "[Talk show host Jay] Leno once told a magazine he had not eaten a vegetable since 1969, and he insisted he tasted his last apple in 1984" (Huffington Post 2012). When Michelle Obama went on his show, she convinced him to nibble on apples dipped in honey from the White House garden, sweet potato fries, and a pizza topped with eggplant, green pepper, and zucchini. To reach another audience that did not tend to watch news or general interest programming, she appeared on *Iron Chef America* as part of a special White House episode that pitted well-known television chef Bobby Flay and White House Executive Chef Cristeta Comerford in competition against two other well-known television chefs, Mario Batali and Emeril Lagasse (Burros 2009c). Michelle Obama selected from plants grown in the White House garden the secret ingredient that was to be used in the meal cooked on the show. The judges for this episode, which reached 1.5 million Americans, were television chef Nigella Lawson, actress Jane Seymour, and Olympic swimmer Natalie Coughlin. Michelle Obama also produced a public service announcement for television, which reached an estimated 200 million viewers.

Obama also uses White House events to carry her message. Here are but two of many examples. Before the Obamas' first official dinner at the White House, she shepherded six top students from a local culinary institute, together with reporters, on a tour of the White House kitchen and showed them what is involved in staging a state dinner. At another event, in 2010, she brought the spouses of the heads of state of 31 countries to the Stone Barns Center for Food and Agriculture to tour the organic restaurant and organic farm located on an 80-acre former cattle farm on the Rockefeller family estate in New York. The visitors toured the organic herb and vegetable gardens and the egg-laying hens, and watched children from local schools learning to cook with organic ingredients from the farm (Burros 2009a, 2010a, b).

6.2.1 Healthy Schools

Chapter 4 is devoted to the history of school lunch policy and includes the relevant legislation from the Obama Administration, so we present only an abbreviated discussion of this plank in the Let's Move platform. Michelle Obama's main activity in this area has been to actively support the reauthorization of the Child Nutrition Act, which was passed in 2010 with the name Healthy, Hunger-Free Kids Act (Pear 2010; also see Chap. 4). Prior to the bill's passage, she lobbied Congress on the need for more fruits and vegetables on school menus. In January 2012 she and USDA Secretary Tom Vilsack together announced the new nutritional standards for school meals that were set out in the legislation, which came into effect in July 2012.

The bill was not without controversy. While it was passed in the Senate by unanimous consent, the House was divided, voting 264-157 in favor. It was the first increase in the federal subsidy of school lunches in 30 years, and it was an expensive piece of legislation. Although most Democrats supported the bill, some were unhappy that a compromise was made with the Republicans to finance the $4.5 billion cost by a cut in the food stamp program (the Supplemental Nutrition Assistance Program). President Obama assuaged the Democrats by promising to find other places to take the funds from. Many House Republicans voted against the bill, some arguing it as an unfunded mandate on the states and local government because the increased federal budget covers no more than half the cost of the higher quality meals. Because the bill regulates prices on lunches served to children from families with income above 185 % of the poverty line (i.e., approximately $41,000), it requires schools to raise lunch prices in many cases; some Republicans argue that this represents a tax on middle-class families. Many Republicans also argued that decisions about school lunches should be left in the hands of local school boards. These were familiar lines of argument for anyone who had followed federal school lunch policy debates over the years.

Where industry opposition to federal regulation of school food was once strong, it was attenuated in this case. Through a Clinton Foundation initiative with the American Beverage Association, major beverage companies including CocaCola Company and PepsiCo had found alternative products such as bottled waters and juices to sell in the schools. The food industry still would have preferred voluntary rules to federal regulation, but at least the industry had 3 years to make its product lines conform to the new nutritional standards.

6.2.2 Remaking the President's Physical Fitness Challenge Program

A second plank in the Let's Move program to end childhood obesity is the improvement of physical education in schools. According to the *Journal of the American Medical Association*, almost all 9-year-olds get enough exercise, but only 31 % of 15-year-olds do. Only 22 % of high school seniors are enrolled in physical education classes, and there are serious questions about how much exercise students actually receive in these classes. Half of all adults in the United States do not achieve the recommended amounts of physical activity, while one-quarter of all adults do not engage in physical activity at all. Although physical exercise is important in fighting obesity, it is perhaps not as important a policy issue as school lunches. The reason is that lack of exercise is not getting worse while obesity is, so it is presumably diet rather than exercise that is the leading contributor in the growth of obesity. There are of course strong reasons to participate in moderate physical exercise, such as reduced incidence of heart disease and high blood pressure; a combination of good diet and exercise helps with weight maintenance.

Before turning to the story of physical activity during the Obama administration, we take a brief look at the history of physical education in America. With increasing urbanization and industrialization in the nineteenth and early twentieth centuries, Americans generally received less physical exercise than they had when most people worked on farms. For example, a U.S. Treasury Department Report on Exercise and Health in 1915 indicated that Americans were becoming too sedentary. Sports became popular in the late nineteenth century, and Teddy Roosevelt served as a role model for the nation as someone vital and physically fit. Nonetheless, one-third of recruits were found unfit for service in the First World War. Legislation after the war mandated better physical fitness programs in the public schools. But as the war became more distant, people lost interest in this issue; and when money became tight during the 1930s, funding for school fitness programs was eliminated. The Second World War was a reprise of the earlier war, with an even larger percentage of the recruits deemed unfit for service. In the late 1940s, University of Illinois scientist Thomas Cureton conducted the first scientific research on physical activity and health (Rice et al. 1958; Welch 1996; Dalleck and Kravitz 2002). On concerns today of fitness of recruits, see Mission: Readiness (2010a, b).

The president's role in this national effort to enhance school physical education comes most directly through the council on physical fitness he appoints. President Eisenhower founded the President's Council on Youth Fitness in 1956 to encourage children to be healthy and active, after he had read a report that American kids were less fit than European kids, when measured by the Kraus–Hirschland muscular fitness tests. Vice President Richard Nixon, who had enthusiastically played basketball and football in college, was appointed as the first chair of the Council. In 1963 President Kennedy changed the name to the President's Council on Physical Fitness, so that its mandate included all Americans and not just youth. President Kennedy was concerned about America's physical fitness and wrote two articles about it for *Sports Illustrated* (Kennedy 1960, 1962). In 1966 President Johnson added "Sports" to the Council's title in recognition of the growing importance of sports in American society. In 1982 President Reagan added to the Council's purview research, amateur sports, and sports medicine. Perhaps the Council had its highest profile under President George H. W. Bush, when he appointed Arnold Schwarzenegger as chair. Schwarzenegger held "American Workouts" on the White House lawn and visited all 50 states to promote physical fitness.

The Obama administration was by no means the first voice calling for better school physical education. The American Medical Association has been calling for strong school physical education programs since 1960. The highest priority in *The Surgeon General's Call to Action to Prevent and Decrease Overweight and Obesity* (U.S. Department of Health and Human Services 2001; Also see U.S. Department of Health and Human Services 1996) was daily, quality physical education in all school grades. The American Academy of Pediatrics has also argued strongly for these programs. Public interest in these issues was further stirred by popular books such as *Fast Food Nation*, *Food Politics*, and *Fat Land* and by the movie *Super Size Me* (Schlosser 2001; Nestle 2002; Critser 2004; Spurlock 2004; Also see Ardell 2003).

Despite all these calls for action, no federal law exists to mandate school physical education. Historically, this policy has been regarded as a state and local issue. A small number of states have set regulations for physical education, especially in the years between 2000 and 2010, but most of their regulations do not specify how much physical education there should be or how much time in a physical education class should be devoted to actual physical activity by the students. In fact, few, if any states meet the national recommendations of 150 min of physical activity per week at the elementary school level and 225 min at the high school level.

One of the major responsibilities of the U.S. Department of Health and Human Services has been to track and promote the physical well-being of the population of the nation as a whole, as well as various subgroups identified by race, ethnicity, age, income, education, or disability. One way of doing this has been through three, 10-year health objectives reports for the nation, each one providing benchmarks to meet in the coming decade—known as *Healthy People 2000*, *Healthy People 2010*, and *Healthy People 2020* (Fulton et al. 2011; U.S. Department of Health and Human Services 2000, 2007, 2010). The Healthy People plans are about more than physical activity. They also consider issues related to health care, education, and public health. For example, *Healthy People 2020* covers 42 topic areas and identifies 1,412

Reducing the number of people with no leisure-time physical activity
Increasing aerobic activity and muscle strengthening
Increasing school requirements for physical education programs
Increasing the number of states and schools districts requiring school recess (where some physical activity occurs)
Reducing hours spent watching television and playing video games
Introducing licensing requirements requiring various kinds of physical activity during child care
Increasing multiple use of physical activity spaces and facilities (such as school tennis courts) for the public outside of regular school hours
Increasing the number of doctor visits that involve counseling or education related to physical activity

Fig. 6.1 Benchmarks associated with the *Healthy People 2020* report

objectives. Some objectives focus on reducing or eliminating illness or disability; others on eliminating health disparities, improving access to health care, or improving dissemination of health information. *Healthy People 2020* is supported by the Healthy People Consortium, a group of several hundred private and public organizations. The Centers for Disease Control and the Council worked with several other agencies to formulate the physical activity component of *Healthy People 2020.*

The United States is not the healthy nation that the organizers of *Healthy People 2010* had called for. Of its 15 physical activity objectives, divided between physical activity behaviors (e.g., increasing aerobic physical activity) and policies to support physical activity (e.g., increasing worksite physical activity programs), the nation met none of these benchmarks. This performance was kept in mind in the creation of the benchmarks for *Healthy People 2020.*

Healthy People 2020 addresses physical activity behavior, environments, and policies. Its benchmarks are presented in Fig. 6.1.

President Obama changed the Council's name in 2010 to the President's Council on Fitness, Sports and Nutrition in order to give greater emphasis to health and less emphasis to numbers of push-ups and sit-ups. He appointed New Orleans Saints quarterback Drew Brees and Olympic gymnast Dominique Dawes as co-chairs. Council appointees include many well-known athletes, but also doctors, chefs, nutrition experts, and leaders of philanthropic foundations interested in health such as the Robert Wood Johnson Foundation. At the meeting to introduce the new members of the Council, Michelle Obama jumped rope and made a pitch for healthy eating and vigorous physical activity (Eyler 2011; also see Payne and Morrow 2009).

President Obama's Council developed a four-part program called The President's Challenge to encourage Americans to be active and physically fit. A Physical Fitness Test measures people's ability to carry out traditional physical tasks such as sit-ups. An Adult Fitness test covers aerobic fitness, muscular strength and endurance, flexibility, and body composition. A Presidential Active Lifestyle Award (PALA) provides recognition to individuals for reaching particular physical activity and healthy eating goals. The Presidential Champions component is intended for people who are already physically fit, as a next step after PALA. To encourage fun and provide incentives to participate, the system is designed so that people receive points for particular achievements and can seek to attain various award levels.

People can participate in the Challenge on their own or they can form their own groups and compete within the group. Thousands of government employees were encouraged to sign up for PALA.

Online tools were developed so that people could track their performance. The website was hacked in early 2012, making some mildly personal information available to the hackers. While the email addresses and date of birth were part of these records, there were no Social Security numbers or credit card information, and even names were optional. It is hard to understand why anyone would want to hack this site, other than that it could be done and that it was associated with the White House. Perhaps the most amusing thing is that when it was determined that the site had been compromised, a notice was posted that until the problem was fixed, the site was "taking a breather."

One can get a sense of the philosophical direction of Obama's Council by reviewing some of the papers in the Council's research digest. Some politicians had questioned the efficacy of policy in the physical activity realm, given how ineffective state and local regulation of school physical activity have been at improving children's physical fitness. However, the Council remains a true believer in the value of policy. It points to the salutary effects of policy in other public health areas such as water fluoridation, safety belt regulation, secondhand smoke, and elimination of *trans*-fats in restaurants. In fact, it sees school physical activity as only one of five ways to use policy to improve the nation's physical fitness:

1. Quality physical education in schools—including increasing the amount, teacher certification, provision of facilities and equipment, and reduction of exemptions from physical education classes for extracurricular activities.
2. Complete streets policy—redesigning communities to include bike lanes and sidewalks as a way to encourage bicycling and walking.
3. Joint use policies—making school gymnasia, tracks, playing fields, and swimming pools available to the public.
4. Community trail policies—for walkers, joggers, equestrians, and off-road bikers.
5. Active transportation policies—increasing opportunities for kids to walk or bike to school.

In addition to jumping rope at the President's Council on physical fitness meeting, Michelle Obama has also exercised in other public forums. For example, she did 25 pushups on Ellen Degeneres's television show, raced Jimmy Fallon around the White House, and participated in a photo-op doing hula hoop revolutions on the South Lawn of the White House for her book *American Grown* (Obama 2012; Dwyer 2012). She also continues to talk about having a healthy life style. For example, she participated in a live web chat sponsored by the National Association for Sport and Physical Education; discussed her daily workout routine on Team Mom, a website run by Shine from Yahoo, that addresses matters of concerns to American moms; and talked with a reporter from the Huffington Post's Healthy Living section about how she rises at 4:30 or 5:00 a.m. most mornings to get in her exercise before her children wake up (National Association for Sport and Physical Education 2012; also see iVillage 2012).

6.2.3 Nutrition Labeling, Food Pyramids, and BMI Monitoring

The third plank of the Let's Move campaign is to give parents better information and better choices about the foods and portions that their children should eat, as well as providing a means for checking to see if one's children are in a healthy body weight range. Because busy parents often take their children out to eat fast food, a major appeal has been made to major fast-food restaurant chains to help make their food more nutritious. The number of fast food restaurants in the United States grew from 30,000 in 1970 to 233,000 in 2004. Since 2010, a group of advisers to Michelle Obama has been talking with the National Restaurant Association to convince restaurants to offer children's meals with smaller portions and more healthy offerings such as carrots, apple slices, and milk instead of French fries and soda. Citing research that shows children consume more saturated fat and less fiber and calcium when they eat out, Michelle Obama called on restaurants to change menus, recipes, and marketing practices so that parents can be confident that kids will get a healthy meal when they eat out (Stolberg and Neuman 2011). There still has been no complete resolution.

Efforts to require restaurants to provide nutritional content information have been mostly unsuccessful. The FDA has authority to regulate food under the Food, Drug, and Cosmetic Act of 1938. The law requires food labels to be truthful and not misleading. The FDA began to expand regulation of food labeling under the Nutrition Labeling and Education Act of 1990. This law requires a nutrition facts label on most food products, providing information about fat, cholesterol, sodium, carbohydrates, and sugar. Restaurants are exempted from this regulation unless they make specific health claims about particular foods on their menu. In 2003, the FDA required product nutrition labels to list *trans*-fat content on their labels within 3 years. In 2005, the FDA proposed that products provide information on serving size and calorie content per serving.

Restaurants have typically not been regulated because of the lack of uniformity in food preparation and serving size. However, fast food restaurants have high uniformity. No federal or state law requires restaurants to give consumers nutritional content information, even though more than a dozen states introduced legislation between 2000 and 2010. This information continued until recently to be voluntary and appeared in many places: menus, websites, brochures, tray liners, food wrappers, and posters. In 2005 only 44 % of the largest 300 U.S. restaurant chains provided nutritional information for their standard menus. Those providing nutrition information most often offered it on their website, where it is hard to access while a consumer is at the restaurant making a selection from the menu. Through a more recent agreement with the state attorneys general, however, today most fast-food restaurants are now giving nutritional information either on posters in their restaurants or on their websites.

There have also been efforts to use the courts to mandate additional nutritional information from restaurants. Between 2002 and 2006, several personal injury lawsuits were filed against food companies for obesity-related health issues.

The best known of these cases is *Pelman ex rel. v. McDonald's Corp.* (2003), in which the plaintiff argued that McDonald's was negligent in informing consumers of the risks of eating its foods and employed deceptive marketing practices. The court disallowed the certification that would have made this a class-action suit, the plaintiff lost the jurisdictional battle to keep the case in state court rather than having it moved to federal court, and the negligence was not proved. Such cases are hard to win because the plaintiff must prove both injury and that the danger is not clear to consumers. Moreover, many states have laws that specifically make fast-food restaurants immune from obesity lawsuits.

Bowing to pressure from health advocates, McDonald's agreed in July 2011 that by April 2012 it would reduce the number of French fries and add a small portion of apple slices (without the caramel dipping sauce) to Happy Meals, as a way to cut calories by 20 %. Parents were also given a choice of more fruit or veggies instead of the fries, and fat-free chocolate milk or low-fat milk could be selected instead of soda. Parents balked at further menu tinkering with children's meals where fries or soda were not an option. McDonald's already offered customers the option of replacing fries with a serving of apples, but only 11 % of parents have elected this option. McDonald's has agreed that by 2015 it will reduce sodium content by 15 % in all foods except sodas and desserts. The Happy Meal, introduced by McDonald's in 1979 and accounting for 10 % of company sales, will still carry its free toy (Strom 2011).

The city of San Francisco passed a law in 2011 that banned inclusion of toys in meals unless certain nutritional standards are met to ensure that a toy is not an enticement to unhealthy eating. The nutritional changes that McDonald's is making do not meet the San Francisco nutritional requirements for a meal with toys, so there had been speculation in the press about whether McDonald's would stop serving Happy Meals there. McDonald's considered challenging the San Francisco law in court, but instead decided to offer Happy Meals without the toy. However, purchasers of a Happy Meal can buy the toy as a separate purchase for an additional charge of 10 cents. Given the pressure placed on parents by their children, the practical outcome of this law is that often the child still gets the toy and McDonald's makes additional revenue off the transaction (Eskenazi 2011).

Other restaurant chains have complied with the spirit as well as the letter of the law. Jack in the Box banned all toys in its meals. Following recommendations from the National Restaurant Association, Burger King, IHOP, and others introduced healthier options for children. Michelle Obama urged Congress to require restaurants to print nutrition info on menus, and this ended up as a part of Barack Obama's health care law.

Let's Move has also engaged in efforts to improve packaging of food goods sold in grocery stores. An Institute of Medicine task force offered recommendation in May 2010, and Michelle Obama decided not to endorse any corporate plan unless it met the Institute recommendations. One attempt has involved getting beverage makers to place calorie and nutrition labeling on all cans. Another effort has urged industry to have the front side of food packages labeled so as to clearly spell out harmful ingredients such as salt, sugar, and fat. However, the food industry has pushed back and wants to ensure that healthy ingredients, such as calcium and fiber, are also labeled on the front of the package, to balance out the list of unhealthy ingredients. The Obama administration found labels that showed both the healthy and unhealthy ingredients were too complex and confusing, Michelle Obama remained silent, and the White House press office issued a tepid statement indicating that the industry response is "a significant first step." Some critics are concerned that Michelle Obama will be co-opted by industry, and that the final changes that result will be insignificant (Neuman 2011c; Fulton and Langlois 2011).

Michelle Obama has supported changes to the basic information supplied by the government to families about food and nutrition. For the past 20 years, the main tool has been the food pyramid (see Fig. 6.2). Nutritionists never liked the pyramid because it was confusing and did not differentiate between healthy foods such as whole grains and fish and less healthy alternatives such as white bread and bacon. The release of the food pyramid was originally delayed because of complaints from the meat and dairy industries that their foods were being stigmatized by being near the top of the pyramid, which visually suggested they should be eaten in smaller amounts. After some negotiations and minor modifications, the pyramid was released in 1992. A revised version, called MyPyramid, was released in 2005. MyPyramid turned the image on its side, presented colored stripes to represent various food groups, and had a stick figure running up the side to indicate the need for exercise. Nobody was happy with the revised version, which created great confusion.

In 2011 the Obama administration replaced the MyPyramid with MyPlate, a plate-shaped symbol with wedges for the basic food groups (fruits, vegetables, grains, protein) and a small adjacent circle (reminiscent of a glass of milk or a cup of yogurt) for dairy items. Half of the plate is portrayed as filled with fruit and vegetables to represent the main government recommendation for a healthy diet. The new plate was developed over a 2-year period, by Robert Post at the USDA Center for Nutrition Policy and Promotion. It cost $2 million to conduct focus groups, develop and promote the logo, conduct research, build the web site, and publicize the image. Michelle Obama was deeply involved in the food pyramid remake, and she was one of the major speakers, together with USDA Secretary Tom Vilsack and Surgeon General Regina Benjamin, when it was unveiled at a press conference in June 2011 (Rolling Out 2011; Neuman 2011a, b). The intention is to give recommended serving size as well as the recommended types of nutritional food. Post believes the new version is more attractive and easier for children to comprehend than MyPyramid. However, it does not seem to have much greater explanatory value, nor does it have much persuasive value. It has received wide criticism (see, for example, Portnoy 2011).

MyPlate is accompanied by a version called MiPlato, which is targeted at Spanish-speaking American families. There is a greater percentage of obesity among Hispanic Americans than Caucasian Americans. Goya Foods, the largest Hispanic-owned U.S. food company, has been generous in helping Michelle Obama to promote MiPlato (Hispanic Business 2012). Goya created a brochure with recipes meeting the USDA dietary guidelines and offering ten tips for healthy eating. Its sales force

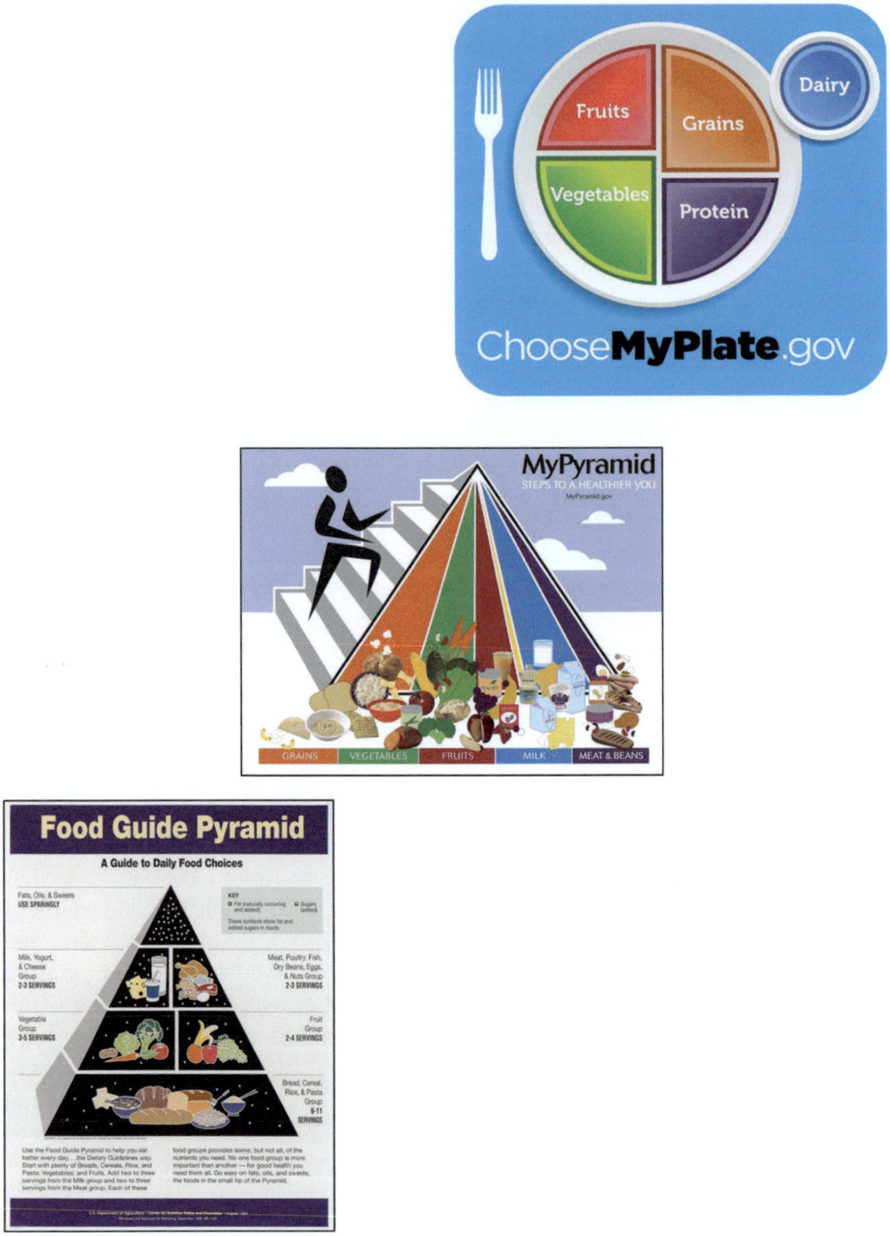

Fig. 6.2 The Food pyramid, MyPyramid, and MyPlate

offers these brochures and MiPlato posters to supermarkets nationwide. Product brochures with a Goya savings coupon are distributed to various national Hispanic organizations at festivals and events that Goya sponsors throughout the year. Goya has created a MiPlato cookbook, which it distributes to its customers. The company has also developed lesson plans for fourth through sixth graders, using beans as educational characters; these are being distributed to elementary schools in six large U.S. cities that have sizable Hispanic populations. The company is also printing the MiPlato image on the packages of six Goya bean products (White House 2012).

Goya and Let's Move are coordinating with various organizations to distribute these materials: local churches, food pantries, community health worker networks, registered dieticians, and schools. The National Latino Evangelic Coalition is distributing these materials at its Nuestro Futuro rallies across the United States. The New Mexico Collaboration to End Hunger is working with state social services and statewide food banks to make sure families receive the materials. The materials are also being distributed through food banks in Texas, Florida, Arizona, New York, and California as part of the Feeding America program. The Girl Scouts organization in Tampa, Florida is promoting MiPlato. Several other organizations have joined as community partners in the USDA Center for Nutrition Policy and Promotion's Nutrition Communicators Network.

Another effort to educate the public involves providing families with guidance about whether their children are healthy. The American Academy of Pediatricians is working with doctors and other health providers to make BMI measurements a regular part of health care screening and to provide parents with guidelines on how to reduce the BMI when the number is too high. When talking about her daughters gaining weight, Michelle Obama reported in 2010 that she had received their BMI scores from the pediatrician but did not know at first how to respond, but she made a few changes such as less television, smaller portion sizes, water in lunch boxes, low-fat milk, apple slices at lunch, and colorful vegetables at dinner that helped bring the BMI into the healthy range (Associated Press 2010). In March 2011 she wrote an op-ed piece in *Parenting* magazine, encouraging parents to get a BMI screening for their kids (Alina 2011).

An example of self-regulation in this area is the Smart Choices Program, a collaboration between the Keystone Center—a nonprofit organization that brokers public–private cooperative solutions to healthcare and other public problems that have a scientific element—food manufacturers, retailers, and public health researchers. The goal is to find a clear, simple, uniform, but voluntary system for labeling the front of food packages. Under the program design, a green-and-white label, together with a checkmark and the words "Smart Choices Program: Guiding Food Choices" will be printed on the front of food packages that meet the health criteria of the program. The criteria, divided into 19 product categories, are based primarily on the USDA's Dietary Guidelines for Americans. The signage is intended to replace the many different proprietary systems used by food manufacturers and retailers to designate their healthy food choices. In addition to the Smart Choices logo, the label includes information about the number of calories per serving and the number of servings per container. It is not clear how rigorous the standards will be for attaining Smart Choice status, and no plan has been made for objective evaluation of the impact of the program. The program is currently suspended until the USDA sets rules about front-of-package labeling. (For more information, see smartchoicesprogram.com.)

6.2.4 Access to Affordable Healthy Food

There was a growing recognition in the 1960s of a national problem with hunger in America, punctuated by such events as the Poor People's March on Washington, DC, organized by Martin Luther King, Jr. and carried out in 1968 soon after his assassination. In 1969 the White House Conference on Food, Nutrition, and Health pointed to the special nutritional needs of pregnant women and small children. In response, that same year the federal government established the Commodity Supplemental Food Program. Unfortunately, that program did not meet the needs, and in 1972 the WIC program was passed as the Special Supplemental Food Program for Women, Infants, and Children. It was based on a food voucher program that had been tested in a neighborhood of Baltimore by Johns Hopkins University. The main goal was to provide nutritional foods at critical times of human growth and development—to serve as a supplement to food stamps. The program was tweaked various times, including a major effort to control costs in the 1980s in light of the high cost of infant formula. Reaching 88,000 people in 1974, 25 years later it served over seven million people. The program is coordinated with public health efforts, and it has been a major pathway for poor families to enter the public health system (USDA 2012, n.d.; Oliveira et al. 2002; Center for Study of the Presidency and Congress 2012).

The food stamp program began in the Unites States in 1936. People on relief could buy orange food stamps at face value and use them for any food, and receive 50 cents worth of blue stamps free for every $1 of orange stamps purchased. Blue stamps could be used to buy from a list of foods determined by the USDA, mostly surplus foods. The program ended in 1943, mainly because there was not much surplus food available during the war and many fewer people were unemployed. The highest number of people using stamps from that program at any one time was four million (MacDonald 1977; New York Times 2010).

After the Second World War ended, various liberal members of Congress proposed a new food stamp program, but no action was taken during the Truman or Eisenhower administrations. Soon after his inauguration in 1961, President Kennedy began a pilot program. It expanded over the next 3 years from eight to 43 regions of the country. In 1964 President Johnson requested that Congress make the food stamp program permanent. The Food Stamp Act passed that same year. Participation expanded from half a million in 1965 to 15 million in 1974, the year in which the program was finally available in every county in the country.

Over the years, there were many fine tunings to the program. The Food Stamp Act of 1977, championed by the Democrats, eliminated the purchase requirement so that a larger number of people could participate. Eligibility was set for families at or below the official poverty line. There were cutbacks in the 1980s, for example, making more types of revenue such as retirement funds count toward family income, placing some people above the poverty line and thus removing their eligibility for food stamps. An electronic benefit transfer system was introduced in 1984 so that the federal assistance could be paid directly to the grocer by debit card. Funding increased in 1993 with passage of the Mickey Leland Childhood Hunger Relief Act, and by 1994, 28 million Americans were using food stamps. The Personal Responsibility and Work Opportunities Reconciliation Act of 1996, part of general welfare reform of the mid-1990s, eliminated food stamps for most legal immigrants and restricted the length of time able-bodied adults without children could receive food stamps. As a consequence of this bill and falling unemployment, participation in the food stamp program dropped steadily during the late 1990s. The Food Security and Rural Investment Act of 2002 reinstated food stamps for legal immigrants, so long as they had been living in the United States for 5 years. Participation began to increase and had grown to 26 million people by 2006. The Food, Conservation, and Energy Act of 2008 increased funding and changed the program's name to the Supplemental Nutrition Assistance Program (SNAP) in hope of reducing the stigma attached to receiving food stamps.

6.2.5 Food Deserts

The last of the four planks in Michelle Obama's Let's Move campaign is the elimination of so-called food deserts, rural and urban communities where there is no access to affordable grocery stores carrying wholesome foods. Food deserts have higher incidence of obesity and low nutrition because people often have to shop in small corner stores that do not stock fresh produce, meat, and dairy items and sell only overpriced, processed foods. (According to the government's working definition, a food desert is a place where at least 500 people or a third of the population lives more than a mile away from an affordable food store. Maps of food deserts can be found in the Food Environment Atlas of the USDA at http://www.ers.usda.gov/data-products/food-environment-atlas.aspx.) Food deserts exist in both inner cities and rural areas. African Americans are four times more likely than whites to live in a food desert.

The newsworthy and somewhat controversial first alliance in Let's Move's effort to eliminate food deserts came through a collaboration with the retailer Walmart. Michelle Obama believed that if she could get Walmart, America's largest grocery chain, to create nutrition labels that are easier to understand; reduce prices on fruits and vegetables; and reduce the amount of sugar, fat, and salt in the food products it sold, other supermarkets, which often use the same suppliers as Walmart, would follow Walmart's practices. Walmart had been interested in opening new stores in inner cities, but it had hesitated in doing so both because of the economic uncertainties and because its right-to-work policies clash with pro-labor sentiments commonly found in inner city communities. Encouraged by Michelle Obama's goals (or some might argue, taking advantage of them), Walmart announced a 5-year plan in 2011 to lower salts, fats, and sugars in many of the products it sells and to lower prices on healthy items. Walmart also agreed to put signs on packages indicating which foods are healthier, address food deserts, and provide more support for nutritional education programs (Bradley 2011).

This arrangement received criticism from the political right, in particular from Rush Limbaugh, who argued that the Administration was bullying a company. It also received criticism from the political left because of Walmart's labor practices. As a Senator, Barack Obama had complained in 2007 about the low wages that Walmart pay, making it more interesting to see his wife collaborating with the company.

While Walmart made its plans to open 300 new stores in inner cities, Michelle Obama partnered with other grocery and drug store chains, and with private foundations, to work on the food desert problem. The stated goal of the Let's Move campaign is to open or expand 1,500 stores over a 5-year period in rural and urban settings. It is expected these efforts will reach 23.5 million people, including 6.5 million children. The largest partnership planned is with the largest U.S. drug store chain, Walgreens, which hopes to expand at least 1,000 of its drug stores into food oasis stores. Michelle Obama also received commitments from the national grocery chain Supervalu and from various regional grocery chains (Rivas 2011).

Chicago has become a major pilot program for the nation. In 2011, Michelle Obama teamed up with Chicago mayor Rahm Emanuel to eliminate food deserts in her hometown. They plan to reduce Chicago's food deserts by 20 % and create 2,000 jobs through a partnership with four grocery chains: Supervalu, Roundy's Supermarkets, Walmart, and Aldi's Store. The plans call for 17 traditional grocery stores and 19 expanded Walgreen's. A local urban farm network will supply locally grown foods to Walgreen's and Aldi's. In addition, the Kraft Foundation and the Safeway Foundation have donated $150,000 to open five new farmer's markets in the midst of Chicago food deserts (Sweet 2011).

California has also turned out to be a major pilot for the food desert effort, through a program heavily promoted by the Let's Move campaign called the FreshWorks Fund. This fund is a partnership between the California Endowment—a Los Angeles-based foundation focused on health issues—and several California-based banks and health organizations to open grocery stores with healthy foods in areas that do not currently have grocery stores because of economic or safety reasons. The fund has provided $20 million to California-based Northgate Gonzalez Markets to build stores in three underserved areas (Inglewood, San Diego, South Los Angeles) (Hoag 2012). Gonzalez, a supermarket chain founded by a Mexican immigrant, operates supermarkets in the Los Angeles, Orange County, and San Diego areas and employs 5,000 people. It helps to promote nutrition in Hispanic communities, not only through healthier food choices but also through cooking classes and educational campaigns. The FreshWorks Fund is also supporting the development of a farmer's market near a public housing project and a food delivery service for outlying rural areas.

President Obama's federal 2011 budget included the Healthy Food Financing Initiative, calling for $400 million over 7 years to open grocery stores in food deserts and providing incentives to convenience stores to offer more healthy options. The administration's goal is to eliminate food deserts within 7 years. A bill to fund this initiative was introduced in 2011 by Rep. Allyson Schwartz (D-PA), but it died in committee. This initiative received $32 million in the megabus appropriations bill for fiscal 2012, approved in December 2011. A total of $22 million goes to the Department of Treasury, which is channeling it to community development financial institutions, which will lend the funds to food retailers serving customers in food deserts. The other $10 million will support work on food deserts through the Health and Human Services Community Economic Development Program. Rep. Schwartz (D-PA) reintroduced the bill in 2013; at the time of this writing, it looks to have little chance of being enacted as law.

6.3 Opposition to Federal Regulation

Despite some bipartisan political support and even support from the food industry, Michelle Obama and the Let's Move campaign have had their detractors on both the left and the right. Critics on the left are worried that she will not be tough enough on the food industry and will settle for changes that are little more than window dressing. For example, these critics want strict bans on junk food in school, not just more nutritious school lunches served. Another approach would be to alter the economy of food subsidies so that junk food is no longer less expensive than nutritious food in America's grocery stores.

Not surprisingly, there has been stronger criticism from the right. As mentioned earlier, some Republicans have argued that the school lunch program is a partly unfunded mandate, and others see the raise in school lunch cost as a tax on the middle class. Sarah Palin received considerable publicity when she brought cookies to a private Christian school in Bucks County, PA (Warner 2010). She argued that the antisugar regulations were taking treats out of the hands of kids at school and that the Michelle Obamas of the world were taking pleasure away from kids and attacking the American way of life. It seemingly did not matter to Palin that neither was this a public school subject to the school lunch laws, nor did the school have any antisugar regulation. Conservative radio host Glenn Beck complained about federal regulation eliminating people's right to choose to be fat. Other conservatives angrily questioned Michelle Obama's right to a bully pulpit and charged her with hypocrisy, as indicated in the D'Andrea epigraph at the beginning of this chapter (Rohter 2011).

For all of its educational and symbolic advantages, there are flaws in the White House garden model. The supposed low cost ($200) of raising a large garden does not take into consideration the extensive free labor, or the added cost of gardening tools and fencing to reduce theft and vandalism. The model also assumes usable soil that might not be available in polluted, vacant city lots. Whether the White House Garden or the Let's Move Initiative will come to fruition in making the United States a more healthy place, it is too soon to tell.

References

Alina (2011) Michelle Obama encourages BMI screening for kids. Show & Tell (blog), Parenting.com. http://www.parenting.com/blogs/show-and-tell/alina-parentingcom/michelle-obama-encourages-bmi-screening-kids. 23 March 2011, Accessed 20 Dec 2013

Ardell DB (2003) It's time to revitalize the President's Council on Physical Fitness and Sports (PCPFS). Don's Report Archive. May 5. http://www.seekwellness.com/wellness/reports/2003-05-05.htm. 5 May 2003, Accessed 20 Dec 2013

Associated Press (2010) Michelle Obama worried about daughters' BMI. http://www.msnbc.msn.com/id/35144040/ns/health-diet_and_nutrition/t/michelle-obama-worried-about-daughters-bmi/#.UKjvTaVBkrh. 29 Jan 2010, Accessed 20 Dec 2013

Bailey H (2009) No White House food fight. Newsweek. 26 Jan 2009, 153(4):10

Batra-Wells P (2010) One nation under arugula: The Obama White House Garden as cultural display and pedagogy. Master's Thesis, Bowling Green State University

Black J (2009) Vegetable garden will be installed on White House grounds. Washington Post. March 20. http://www.washingtonpost.com/wp-dyn/content/article/2009/03/19/AR2009031902886.html. 20 March 2009, Accessed 20 Dec 2013

Bradley T (2011) Michelle Obama and Walmart join forces promoting healthy food. ABC News. http://abcnews.go.com/Politics/WorldNews/michelle-obama-walmart-join-forces-promote-healthy-eating/story?id=12723177#.T6qBvO2aErg. 29 Jan 2013, Accessed 20 Dec 2013

Burros M (2009a) Before big dinner, the First Lady gives a kitchen tour. New York Times. http://www.nytimes.com/2009/02/23/us/politics/23dinner.html. 22 Feb 2009, Accessed 20 Dec 2013

Burros M (2009b) Obamas to plant vegetable garden at White House. New York Times. http://www.nytimes.com/2009/03/20/dining/20garden.html?ref=politics. 19 March 2009, Accessed 20 Dec 2013

Burros M (2009c) Someone's in the kitchen with Michelle: The secret ingredient is politics. New York Times. http://www.nytimes.com/2009/11/04/dining/04iron.html. 3 Nov 2009, Accessed 20 Dec 2013

Burros M (2010a) Promoting healthy eating, with 31 other first ladies. New York Times. http://www.nytimes.com/2010/09/25/nyregion/25firstlady.html. 24 Sept 2010, Accessed 20 Dec 2013

Burros M (2010b) Healthful food takes no holiday at the White House. New York Times. http://www.nytimes.com/2010/12/01/dining/01white.html. 30 Nov 2010, Accessed 20 Dec 2013

Burros M (2012) Michelle Obama reveals how her White House Garden grows. New York Times. http://www.nytimes.com/2012/05/29/us/politics/michelle-obama-writes-american-grown.html. 29 May 2012, Accessed 20 Dec 2013

Cameron CW (2012) Presidential dining. Atlanta Journal-Constitution. 16 Feb 2012, p 1F

Center for the Study of the Presidency and Congress (2012) The history of SNAP. SNAPtoHealth.org. http://www.snaptohealth.org/snap/the-history-of-snap. Accessed 16 No 2012

Critser G (2004) Fat land: How Americans became the fattest people in the world. Houghton Mifflin, Boston

Dalleck LC, Kravitz L (2002) The history of fitness. IDEA Health and Fitness Source. http://www.ideafit.com/fitness-library/the-history-of-fitness. Accessed 20 Dec 2013

D'Andrea H (2012) Food police chief Michelle Obama unveils new federal government nutrition rules all schools will have to meet. AZPundit (blog). January 26. http://azpundit.com/food-police-chief-michelle-obama-unveils-new-federal-government-nutrition-rules-all-schools-will-have-to-meet. 26 Jan 2012, Accessed 20 Dec 2013

Dwyer D (2012) Michelle Obama does 25 push-ups. Political Punch (blog), ABCNews.com. February 2. http://abcnews.go.com/blogs/politics/2012/02/michelle-obama-does-25-push-ups. 2 Feb 2012, Accessed 20 Dec 2013

Eskenazi J (2011) Happy Meal ban: McDonald's outsmarts San Francisco. The Snitch (blog), SF Weekly.http://blogs.sfweekly.com/thesnitch/2011/11/happy_meal_ban_mcdonalds_outsm.php. 29 Nov 2011, Accessed 20 Dec 2013

Eyler AA (2011) Promoting physical activity through policy. Research Digest of the President's Council on Fitness, Sports & Nutrition. http://www.fitness.gov/pdfs/research-digest-september-2011.pdf. Accessed 20 Dec 2013. 12(3)

Fournier R (2011) No TR: The limits of Obama's bully pulpit. Decoded (blog), National Journal. http://decoded.nationaljournal.com/2011/12/obama.php. 6 Dec 2011, Accessed 20 Dec 2013

Fulton A, Langlois M (2011) Food industry and health experts face off over package labeling. National Public Radio. http://www.npr.org/blogs/health/2011/06/22/137344357/food-industry-and-health-experts-face-off-over-food-package-labeling. 22 June 2011, Accessed 4 Dec 2013

Fulton JE, Wargo J, Loustalot F (2011) Healthy people 2020: physical activity objectives for the future. Research Digest of the President's Council on Fitness, Sports & Nutrition. http://www.fitness.gov/pdfs/research-digest-june-2011.pdf. Accessed 20 Dec 2013. 12(2)

Gavin P (2012) Michelle Obama ranks on power food list. Politico.com. http://www.politico.com/news/stories/0112/71666.html. 19 Jan 2012, Accessed 20 Dec 2013

Givhan R (2010) Michelle Obama: 'Let's Move' and work on childhood obesity problem. Washington Post. http://www.washingtonpost.com/wp-dyn/content/article/2010/02/09/AR2010020900791.html. 10 Feb 2010, Accessed 20 Dec 2013

Hispanic Business (2012) Michelle Obama teams with Goya Foods to bring 'MiPlato' to Hispanic families. HispanicBusiness.com. January 26. http://www.hispanicbusiness.com/2012/1/26/michelle_obama_teams_with_goya_foods.htm. 26 Jan 2012, Accessed 20 Dec 2013

Hoag C (2012) Michelle Obama: First Lady to promote healthy food for Northgate Gonzalez Markets in California, primarily focused on low-income neighborhoods, Hispanic immigrants. Associated Press. http://www.huffingtonpost.com/2012/02/01/first-lady-to-promote-healthy-food-in-california_n_1246467.html. 1 Feb 2012, Accessed 20 Dec 2013

Holecko C (2013) Michelle Obama Quotes on Childhood Obesity. About.com. http://familyfitness.about.com/od/motivation/a/michelle_obama_quotes.htm. Accessed 14 Dec 2013

Holecko C (nd) Let's Move: Michelle Obama's obesity awareness campaign. About.com, Family Fitness. http://familyfitness.about.com/od/childhoodobesity/a/michelle_obama_obesity.htm. Accessed 7 May 2012

Huffington Post (2012) Michelle Obama's low-calorie school lunches slammed by 'hungry' high schoolers (Video). Huffington Post Blackvoices. http://www.huffingtonpost.com/2012/09/26/michelle-obamas-low-calorie-school-lunch-video_n_1914394.html. 2 Oct 2012, Accessed 5 Dec 2013

iVillage (2012) Michelle Obama on exercise and getting her family moving. Team Mom (blog). http://shine.yahoo.com/team-mom/michelle-obama-exercise-getting-her-family-moving-stressor-210700374.html. 22 Aug 2012, Accessed 20 Dec 2013

Kantor J (2007) Where the votes are, so are all those calories. New York Times. http://www.nytimes.com/2007/11/23/us/politics/23food.html?pagewanted=all. 23 Nov 2007, Accessed 20 Dec 2013

Kennedy JF (1960) The soft American. Sports Illustrated. 26 Dec 1960, pp 15–17

Kennedy JF (1962) The vigor we need. Sports Illustrated. 16 July 1962, pp 12–15

Kessler D (2009) The end of overeating. Rodale, New York

Lynn LE Jr (2009) Theodore Roosevelt redux: Barack Obama confronts American bureaucracy. Int J Public Adm 32(9):773–780

MacDonald M (1977) Food stamps: An analytical history. Soc Sci Rev 51(4):642–658

McCarvel B (2009) MACA letter to White House about organic garden. CropLife.com. http://www.croplife.com/article/1984/maca-letter-to-white-house-about-organic-garden. 31 March 2009, Accessed 20 Dec 2013

Mission: Readiness (2010a) Too fat to fight: Retired military leaders want junk food out of America's schools. http://cdn.missionreadiness.org/MR_Too_Fat_to_Fight-1.pdf. Accessed 20 Dec 2013. Mission: Readiness, Washington DC

Mission: Readiness (2010b) Unfit to fight. http://missionreadiness.s3.amazonaws.com/wp-content/uploads/Unfit-to-Fight-Report.pdf. Accessed 20 Dec 2013. Mission: Readiness, Washington DC

National Association for Sport and Physical Education (2012) First Lady Michelle Obama's live web chat on Let's Move. http://www.aahperd.org/naspe/about/announcements/ask-michelle-obama.cfm. Accessed 19 Nov 2012

Nestle M (2002) Food politics: How the food industry influences nutrition and health. University of California Press, Berkeley

Neuman W (2011a) Goodbye to the food pyramid, hello to the dinner plate. New York Times. May 27. http://www.nytimes.com/2011/05/28/health/nutrition/28plate.html. 27 May 2011, Accessed 20 Dec 2013

Neuman W (2011b) Nutrition plate unveiled, replacing food pyramid. New York Times. June 2. http://www.nytimes.com/2011/06/03/business/03plate.html?_r=0. 2 June 2011, Accessed 20 Dec 2013

Neuman W (2011c) Food makers devise own label plan. New York Times. http://www.nytimes.com/2011/01/25/business/25label.html?_r=0. 24 Jan 2011, Accessed 4 Dec 2013

New York Times (2010) A history of food stamps use and policy. http://www.nytimes.com/interactive/2010/02/11/us/FOODSTAMPS.html. 11 Feb 2010, Accessed 20 Dec 2013

Obama M (2012) American grown: The story of the White House Kitchen Garden and gardens across America. Crown, New York

Oliveira V, Racine E, Olmsted J, Ghelfi LM (2002) The WIC program: Background, trends, and issues. Food Assistance and Nutrition Research Report No. FANRR-27, USDA Economic Research Service. http://www.ers.usda.gov/publications/fanrr-food-assistance-nutrition-research-program/fanrr27.aspx. Accessed 15 Nov 2012

Payne VG, Morrow JR Jr (2009) School physical education as a viable change agent to increase youth physical activity. Research Digest of the President's Council on Physical Fitness and Sports. https://www.presidentschallenge.org/informed/digest/docs/june2009digest.pdf. Accessed 20 Dec 2013. 10(2)

Pear R (2010) Congress sends Child Nutrition Bill championed by Mrs. Obama to President to sign. New York Times. http://www.nytimes.com/2010/12/03/us/politics/03child.html. 3 Dec 2010, Accessed 20 Dec 2013

Portnoy H (2011) Michelle Obama's MyPlatefood guide a spectacular failure. The Greenroom (blog), Hot Air. http://hotair.com/greenroom/archives/2011/12/21/michelle-obamas-myplate-food-guide-a-spectacular-failure. 21 Dec 2011, Accessed 20 Dec 2013

Rice EA, Hutchinson JL, Lee M (1958) A brief history of physical education. Ronald Press, New York

Rivas J (2011). Michelle Obama: Major chains can make healthy foods affordable. ColorLines. http://colorlines.com/archives/2011/07/first_lady_makes_agreements_with_walgreens_and_walmart_to_turn_food_deserts_in_to_food_oasis.html. 25 July 2011, Accessed 20 Dec 2013

Rohter S (2011) Is Michelle Obama turning into a modern day Marie Antionette? Food, diet, and the public health. LessGovistheBestGov.com. http://www.lessgovisthebestgov.com/Michelle-Obama-Food-Diet-Health.html. Accessed 20 Dec 2013

Rolling Out (2011) Michelle Obama supports USDA's change from food pyramid to plate. RollingOut.com. http://rollingout.com/politics/michelle-obama-supports-usdas-change-from-food-pyramid-to-plate. 5 June 2011, Accessed 20 Dec 2013

Salter C (2010) What's cooking in America's schoolyard gardens? Geogr Teach 7(2):56–64

Schlosser E (2001) Fast food nation: The dark side of the all-American meal. Houghton Mifflin, Boston

Spurlock Morgan (dir) (2004) Super Size Me. Distributor: Samuel Goldwyn Films Roadside Attractions

Stolberg SG (2011) First Lady taking health work global. New York Times. http://www.nytimes.com/2011/02/09/us/politics/09michelle.html?partner=rss&emc=rss. 8 Feb 2011, Accessed 20 Dec 2013

Stolberg SG, Neuman W (2011) Restaurant nutrition draws focus of First Lady. New York Times. http://www.nytimes.com/2011/02/07/us/politics/07michelle.html?pagewanted=all. 7 Feb 2011, Accessed 20 Dec 2013

Strom S (2011) McDonald's trims Its Happy Meal. New York Times. http://www.nytimes.com/2011/07/27/business/mcdonalds-happy-meal-to-get-healthier.html. 26 July 2011, Accessed 20 Dec 2013

Swarns RL (2009) Michelle Obama's agenda includes healthful eating. New York Times. http://www.nytimes.com/2009/03/11/dining/11lady.html?pagewanted=all. 10 March 2009, Accessed 20 Dec 2013

Sweet L (2011) Rahm Emanuel, Michelle Obama announce grocery stores for Chicago 'food deserts.' The Scoop from Washington (blog), Chicago Sun-Times. October 25. http://blogs.suntimes.com/sweet/2011/10/rahm_emanuel_michelle_obama_an.html. 25 Oct 2011, Accessed 20 Dec 2013

Todd A (2009) Green thumbs up: Interest in community gardens sprouting all across America. Rural Cooperatives. May/June, pp 26–27

U.S. Department of Agriculture (2012) "Supplemental Nutrition Assistance Program: A short history of SNAP." Food and Nutrition Service. http://www.fns.usda.gov/snap/rules/Legislation/about.htm. Accessed 16 Nov 2012

U.S. Department of Health and Human Services (1996) Physical activity and health: A report of the Surgeon General. Centers for Disease Control and Prevention. http://www.cdc.gov/nccdphp/sgr/index.htm. Accessed 22 Oct 2013

U.S. Department of Agriculture (nd) From food stamps to the Supplemental Nutrition Assistance Program: Legislative timeline. http://www.fns.usda.gov/snap/rules/Legislation/timeline.pdf. Accessed 16 Nov 2012

U.S. Department of Health and Human Services (2000) Healthy People 2010. http://www.healthypeople.gov/2010/. Accessed 22 Oct 2013

U.S. Department of Health and Human Services (2001) The Surgeon General's call to action to prevent and decrease overweight and obesity. http://www.surgeongeneral.gov/library/calls/obesity/index.html. Accessed 18 Sept 2013

U.S. Department of Health and Human Services (2007) Healthy people 2000 Jones and Bartlett learning. [The version includes the midcourse review and the 1995 revisions. The report was originally published in December 1990.]

U.S. Department of Health and Human Services (2010) Healthy People 2010. http://healthypeople.gov/2020/default.aspx. Accessed 22 Oct 2013

Warner J (2010) Junking junk food. New York Times. http://www.nytimes.com/2010/11/28/magazine/28FOB-wwln-t.html. 28 Nov 2010, Accessed 20 Dec 2013

Welch PD (1996) History of American physical education and sport. Thomas, Springfield

Westfall SS (2009) Kitchen cabinet. People. 16 Nov 2009

White House (2012) First Lady Michelle Obama joins Goya Foods in announcing 'Mi Plato' resources for families. Press release, Office of the First Lady. http://www.whitehouse.gov/the-press-office/2012/01/26/first-lady-michelle-obama-joins-goya-foods-announcing-mi-plato-resources. 26 Jan 2012, Accessed 20 Dec 2013

MIX
Papier aus verantwortungsvollen Quellen
Paper from responsible sources
FSC® C105338

If you have any concerns about our products,
you can contact us on
ProductSafety@springernature.com

In case Publisher is established outside the EU,
the EU authorized representative is:
Springer Nature Customer Service Center GmbH
Europaplatz 3, 69115 Heidelberg, Germany

Printed by Libri Plureos GmbH
in Hamburg, Germany